DATE DUE		
JUN 14 1980	MAY 22 1982	
AUG 1 1980	OCT 0 9 1992	
		MAY 25 1993
AUG 7 1982	JUL 30 1993	
JUN 9 1983	MAR 12 1994	
AUG 23 1983	MAR 11 2002	
DEC 10 1983	OCT 16 2008	
OCT 14 1986		
MAR 1 1986		
AUG 16 1988		
FEB 28 1991		
JUN 8 - 1991		

641.872   29,186
J

Jagendorf, M.A.
FOLK WINES, CORDIALS AND BRANDIES

Community Library
Ketchum, Idaho

This book is dedicated to

J O S E P H   D O N O N

great Chef de Cuisine *and matchless connoisseur of wines, in the hope of introducing him to a new-ancient, gustatory-vinous pleasure*

some love, reminiscences, and

THE VANGUARD PRESS, INC.
New York

FOLK WINES,
CORDIALS, & BRANDIES

Ways to make them, together with wise advice for enjoying them

M. A. Jagendorf

with an Introduction by ANDRÉ SIMON

Copyright, ©, 1963, by The Vanguard Press, Inc.
Copyright, ©, 1963, by M. A. Jagendorf.
Published simultaneously in Canada by the Copp Clark Publishing Company, Ltd., Toronto.
No portion of this book may be reproduced in any form without the written permission of the publisher, except by a reviewer who may wish to quote brief passages in connection with a review for a newspaper or magazine.
Manufactured in the United States of America by H. Wolff.
Library of Congress Catalogue Card Number: 63-21854

Picture research and selection by Balance House, Flemington, New Jersey.
Designed by MARSHALL LEE

 INTRODUCTION
by André Simon

WHERE THERE IS NO WATER, there is no life. This is one of the few rules without exception. The grass of the fields and the trees of the forests, like the smallest of insects and the largest of elephants, all must have water to drink if they are to live, and so must you and I. But since the beginning of Time there has been a difference between man and all the other teeming life in and upon and above the earth: other animals expect and get only water, whereas man—and, of course, man embraces woman—throughout the centuries has given a great deal of his hard-earned money to obtain something better than water for himself and for his guests.

Whether or not the countless numbers of drinks, alcoholic or non-alcoholic, invented in all sorts of different countries by all sorts of different peoples, are really better for them than water, matters very little or not at all. All reasonable people know—without having to be reminded of it all the time—that their days are numbered, and they are the more anxious to find the food and the drink and the mate and the sport most likely to add joy rather than years to their lives. Of course, life is all a gamble, and some of us are lucky while others are not so lucky—but luck, like taste, is a matter of personal viewpoint.

Let us remember that we all have different fingerprints, and we shall not be surprised if we also have tastes that are quite different. It is only too true that there are people who cannot experience the joy of taste, just as there are others who have the misfortune to be blind or deaf; however, the great majority of men and women are born with normal senses, that they may know and enjoy what is best to see, hear, eat, and drink. Taste, like any other sense, is subject to the influence of tradition and training.

This is why there have always been, and still are, many people willing to take considerable trouble and to devote a great deal of time and skill to preparing and perfecting drinks made from all kinds of flowers, fruits, and vegetables. The recipes for making these many beverages, as well as their folklore background, will be found in this erudite work of Moritz Jagendorf. He is indeed a well-informed and enthusiastic master.

CONTENTS

INTRODUCTION *by André Simon*

THANKS, 11

PREFACE, 14

"IN THE BEGINNING," 23

AESTHETICS OF WINE MAKING, 28

COMMON SENSE, 31

WHEN, 33

WHAT YOU NEED, 35

    Crocks, 35——*Glass jars*, 38——*Funnels*, 38——*Spoons*, 38——*Straining cloths or bags*, 39——*Scales*, 39——*Press*, 39——*Hydrometer*, 40

THE LIVING FRUITS AND FLOWERS, 41

    Recipes, 41——*Quality and treatment*, 42——*Quantity*, 43——*Increasing the amount of wine*, 43

THE MYSTERY OF WINE, 45

> The yeast, 45——Using the yeast, 48——The vinous fermentation, 50——Sugar syrup, 53

HIGH ADVENTURE, 55

> Sweet and dry, 55——Raisins, 56——Taste and flavor, 57——How strong? 59——Citrus fruits, 59——Heavy and light wines, 60

THE WORSHIP OF BEAUTY, 62

> Straining and fining, 62——Racking, 66

BOTTLING AND THE PLEASURES FOLLOWING, 68

> The bottles and the first law, 68——Corks and corking, 69——Labeling, 71——Record book, 72

THE FINAL WORK, 73

> Maturing and storing, 73——Serving wines, and when to drink them, 77——Of glasses, 79

FRUIT WINES

Apple Wine, 83
Apricot Wine, 95
Banana Wine, 98
Cherry Wine, 100
Date Wine, 104
Grape Wine and Grape-Leaf Wine, 108
Grapefruit Wine, 114
Lemon Wine, 118
Muskmelon (Cantaloupe) Wine, 120
Orange Wine, 122
Peach Wine, 126
Pear Wine, 130
Pineapple Wine, 132
Plum Wine, 136
Pomegranate Wine, 141
Prune Wine, 145
Quince Wine, 147
Raisin Wine, 151
Rhubarb Wine, 155

FLOWER WINES

Carnation (Gillyflower) Wine, 161
Clover-flower Wine, 164
Cowslip Wine, 166
Daisy Wine, 172
Dandelion Wine, 177
Elder-flower Wine, 185
Goldenrod Wine, 189
Marigold Wine, 190
Pansy Wine, 196
Rose Wine, 201

CEREAL & VEGETABLE WINES

Barley Wine, 215
Beetroot Wine, 218
Carrot Wine, 221
Celery Wine, 224
Onion Wine, 226
Parsnip Wine, 230
Potato Wine, 232
Rice Wine (Sake), 238

Spinach Wine, 242
Tomato Wine, 244

Wheat Wine, 247

HERB WINES

Angelica Wine, 251
Balm Wine, 254
Caraway Wine, 258
Chickweed Wine, 261
Clove Wine, 262
Dill Wine, 266

Ginger Wine, 268
May Wine, 270
Mint Wine, 272
Parsley Wine, 276
Sage Wine and Pineapple-sage Wine, 280

BERRY WINES

Blackberry Wine, 285
Blueberry Wine, 291
Cranberry Wine, 293
Currant Wine, 294
Elderberry Wine, 296

Gooseberry Wine, 303
Huckleberry Wine, 305
Mulberry Wine, 307
Raspberry Wine, 311
Strawberry Wine, 315

OTHER WINES

Almond Wine, 321
Bees Wine, 325
Brandy Wine, 327
Bread Wine and Kwas, 328
Cider Wine and Champagne, 332
Honey Wine (Mead), 333

Nut-leaf Wine, 342
Oak-leaf Wine, 346
Rose-geranium Wine, 349
Rose-hip Wine, 352
Sack Wine, 353
Tea Wine, 354

Flowered, Fruited, and Flavored
BRANDIES, LIQUEURS, & CORDIALS

By way of beginning, 361
Neutral spirits, 363
Flaming the flavor, 364
Almond Cordial (Noyau), 365
Angelica Liqueur, 366
Anisette Liqueur, 368
Arak (Araku) Liqueur, 368
Blackberry Cordial, 369
Caraway Liqueur (Kümmel), 369
Carnation (Gillyflower) Cordial, 370
Cherry Bounce, 371
Crème de Cacao, 373
Curaçao, 373
Many-Flowers Brandy: "The Divine Cordial," 374
All-Fruit Brandy, 375
Ginger Liqueur, 375
Lemon Cordial, 376

Marigold Cordial, 376
Mint Cordial, 378
Orange Liqueur, 379
Peach Brandy, 379
Peach Liqueur, 380
Peach, Plum, and Apricot Liqueur: "London's Admirable," 380
Peach and Rose-Petal Liqueur: "Zahedan Liqueur," 381
Pineapple Brandy, 382
Raspberry Brandy, 382
Rose Cordial, 383
Rose Liqueur, 383
Scottish Liqueur, 384
Strawberry Cordial, 385
Strawberry Brandy, 386
Vanilla Liqueur, 386

UNUSUAL DRINKS

Atholl Brose, *389*
Bainnecor Liqueur (Milk Liqueur), *390*
Brown Betty, *391*
Bumpo, *392*
Milk Punch, *392*
Negus, *392*
Ratafia, *394*
Shrub, *394*
Thandai, *396*
Usquebaugh, *397*
White Whey Wine, *398*

TO HELP YOU, 399

Agricultural Colleges and Societies; Garden Clubs and Botanical Gardens, *399*——Special yeasts, *400*——General home-wine-making supplies, *401*——Dried herbs, *401*——Fresh herbs, *402*——Essences, *402*——Books, *402*——Permits, *403*

GLOSSARY, 405

BIBLIOGRAPHY, 407

INDEX, 411

Some Notes about the Illustrations

Facing page 16: An Egyptian queen offering wine to the god Horus. *Following page:* A Bacchanal from a Greek amphora. *Page 25:* Egyptians pressing fruit to make wine. *Page 37:* The press at top is used for making rice wine. *Facing page 65:* The Spanish way of drinking wine from leather bottles, by Doré. *Page 75:* The wine cooler at bottom belonged to George Washington. *Facing page 80:* Left, early Egyptian wine jar; right top, early port bottle; below, English leather bottle. *Following spread:* left, fourteenth-century French tin wine cup; left top, fourth-century wine glasses; bottom, Anglo-Saxon glass; right, wine glasses, Pasco crystal. *Facing page 81:* left, decanter, Pasco crystal; right top, Syrian glass decanter, fourteenth-century. *Facing page 224:* Etching by Manzu, courtesy Museum of Modern Art, New York. *Page 237:* This is the first printed picture of a potato. It appeared in the second edition of Écluse's *Rariorum Plantarum Historia,* 1601.

Much gratitude for their generous assistance is due to the staffs of the Print Room and the Picture Collection of the New York Public Library, the Rare Book Room of the New York Academy of Medicine Library, and the Library of the New York Botanical Gardens. The illustrations on pages 61, 175, 203, 237, 313, and 343 are reproduced by special permission of the Spencer Collection of the New York Public Library.

 THANKS

THERE ARE SO MANY FOLKS in our land and in other lands who had a finger in the pie and a drop in the bottles of this book that I cannot enumerate them all. And so, besides those I mention, there will be many whom I do not, but to whom I am deeply grateful for their kind assistance in creating this volume.

I must begin with gratitude to my good son, André-Tridon, for his invaluable assistance in mysteries of the biochemical action of wine creation. Then there is my good wife, who bears the largest burden of the chores—correction of the manuscript and ideas, etc. Of course, I get all the credit. I hope the Lord will give her ample credit, which she fully deserves, and reward her with a fine herb garden without my ever-bellowing interference.

High on the list also stand H. S. Latham and Dr. J. T. Shipley, both of whom labored valiantly correcting and combing for errors and strayings.

Then there is Miss Elizabeth C. Hall, my botanical angel, librarian of the New York City Botanical Garden, who is ever ready to give her valuable time in research. Thanks are also expressed to all the other members of the library staff. Professor G. Levine, of the Cornell University Agricultural Department, must also be thanked for helping me generously.

Then come a host of good folks from all over, each of whom contributed a part to the book. There was Mrs. Gaskell of Ceylon and Mr. Gill of India and Mr. T. T. Li of the Chinese Mission to the United Nations. There was Branco Novakovic of Yugoslavia and W. P. Springle of England and Mrs. Currie of The National Federation of Women's Institutes of England.

In our own land, Mrs. Bertha Nathan of Maryland and Gene Wagner of Texas both did yeoman's share. Helen Winston of Vermont; Blanche Keysner of West Virginia; Ann Grimes and Dave Webb of Ohio all helped; and so did Mrs. William Hart of Maryland. Then there was Miss Judith Ginsburg, librarian at the Jewish Theological Seminary, and Harry Starr, that most accomplished of scholars, both of whom contributed their knowledge and wisdom. Robert Arbib, with his knowledge of herbs and flowers, helped me a goodly amount.

Nearer home, Mrs. Marion Duhrels and Horace Hillery, that excellent Putnam County historian, did their share.

Even nearer were Bill Harkins, right on my farm, and all the Cole clan (on whose farm I live now), and Vasily Le Gros of the nearby Russian Orthodox Hermitage, all of whom helped and aided with ideas and deeds.

Dr. Margaret Bryant and many more of the New York City and Cooperstown confraternity of folklorists also contributed suggestions and advice. And, finally, there were organizations and their representatives who greatly assisted me whenever I asked for aid.

The staff of Fleischmann's Yeast Company was always ready to help me. So was Mr. J. Fessler of the Berkeley Yeast Laboratory in California. And so were Mr. G. W. Glenn of the West Company in Pennsylvania; Mr. W. White of the Green Drug and Chemical Corporation of Lenoir, North Carolina; and The Wine Institute of California. I must add that there are always some workers in the state agricultural departments ready with assistance. I owe thanks to all these I mentioned and to the many others I did not mention. They all helped unstintingly.

 PREFACE

WHY A BOOK ABOUT folk wines—wines made by folks from recipes handed down by generation after generation? "Providence, in its great goodness, at an early period of the World, considered it necessary to bless mankind with a liquor whose salutary properties would give health and strength to their bodies, exhilarate their minds, and support them with fortitude to execute their labors. Of all the various liquors for the sustenance of human nature, surely none is equal to inspire the soul with hilarity as the LIQUOR DIVINE." I should stop with these wise words of P. P. Carnell, Esq., F.H.S., etc. . . . But there are still one thousand and one more reasons for this book, of which I shall give but three, a right and proper number, for it is the most prevalent in lore and in religion, even as wine itself.

First, wine making is perhaps the oldest occupation of man.

Wine making and wheat growing were probably the first expressions of the awakening civilization in the vast, hazy space of early years. There is a wondrous wealth of lore and literature about viniculture, equal almost to the lore and literature of love. Throughout the ages, men and women made wine in their homes, from grapes, fruits, and cereals, for health and for pleasure.

> "To cleanse the cloudy front of wrinkled care,
> And dry the tearful sluices of despair."

Second, there is a joy in creating, with the labor of your own hands, rare gustatory pleasures that give the creator a sense of omnipotence.

Third, wine making and drinking bring a genial enjoyment that may be classed as one of the great pleasures of life.

> "Wine works the heart up, wakes the wit;
> There is no cure 'gainst age, but it."

sang the poet Pope.

I own a stone-fenced farm of rich land in Carmel, New York. There are many acres of meadows and woods, fruit trees, berry bushes, and endless flowers. Alas, this is perhaps the last farm in the region with deep-bellied, full-uddered cattle. Also, here there are still some old-timers, men and women who have labored around farms. It is my good fortune to have made friends with these folks. I have gone to old-fashioned square dances, to parties, and to communal gatherings, and I have listened to their words. Thus I learned their tales, handed down from father to son, and their way of life—including making wines at home.

Then it came back to me that when I was a boy in Austria, my mother and many of her friends also made wine in their homes, to be drunk on holidays and special occasions. I recalled scenes of soft-yellow candles in brass candlesticks on stiff white tablecloths, with glistening white plates on which fresh fish was served. These were brought in the early morning from Galaz in nearby Rumania, or from the wide-running river Pruth flowing outside our city. Next to the plate with the fish there always stood a little glass filled with sweet, white kümmel wine, a careful product of my mother's hands.

I also remembered the teeming market place where I bought half a penny's worth of kwas, a pleasant, cool, sourish drink, with apples floating on top, made by the Ruthenian peasants from bread.

Then I realized that I, too, had grown up in a life where wine was made at home, just as it was made by my neighbors at Carmel—the Coles, the Cranes, and the Harkinses, all rich with sons and daughters.

The Coles came to America shortly before 1730, the Cranes earlier. There are quite a few tales told about a Captain Crane during the Revolutionary days. Belden Crane, my neighbor, spoke little but to the point. He had a rare love—no, worship—for trees, and no tree could be cut on his "more or less" two-hundred-acre farm during his lifetime.

And so I decided to call this: *Folk Wines, Cordials, and Brandies,* for it tells of the wines and liqueurs folks have made through many marching years. They were always made in an accepted tradition, according to recipes handed down by word of mouth through the sweeping generations. Sometimes they were set down in letters or in old almanacs and diaries. Often they appeared in old herbals and cookbooks—some of them written down thousands of years ago. All these were my sources.

I was initiated into the art of wine making with wild grapes and elderberries by "Clint" Cole, who loved his wine, perhaps a little too well, and also by Mary and John Cole. Clint was fond of relating incidents of the farm that often twined about wines made at home. I began asking questions and learned much. Wild grapes and berries were gathered in 25-pound baskets and brought in by horse

and wagon. Particularly by "Hen" B. Cole (from whose widow, Mary, I bought my farm).

One day Clint spun: "Twelve fellows came to Hen B. You know, he was kind o' sellin' wine in them Prohibition days. They sat around the porch where we are sittin' now, drinkin' elderberry wine. They just kept on drinkin' and drinkin'. An'! Late that night nine of 'em found it kind o' hard gettin' home, an' three, they slept right here on the porch."

In those days, wines from wild grapes and berries were made in twenty- and fifty-gallon barrels, and folks drank them.

Bill Harkins, with the disposition of a Saint Francis of Assisi, told me how to make dandelion wine.

A new world was opened to me. Home viniculture! I had been a devotee of good wine for many years. My father, a rare gourmet, first introduced me, as a youngster, to light French wines with haunting flavors. We would drink these while eating oysters, away from home, for my mother did not approve. Then I had a young love, Ethel, whose Romanian parents were heavy eaters and magnificent drinkers. Never was cloth laid on the table without a pitcher of thick red wine next to the bread. It was continuously replenished from gallon jugs, ever full. Those were wines pure and rich in body and aroma, homemade by Italian immigrants.

Having learned the pleasure of wine drinking, why not acquire the pleasure of wine making! I found a world of new wines to be made from berries and fruits of the trees and earth. Flowers came later.

My first efforts, as I said, were with wild grapes and elderberries. That was in 1945. My son, André-Tridon, did the actual work. I supervised, according to the teachings of Clint Cole and Belden Crane. We made it in a ten-gallon barrel. The next wine I made from dandelions according to Bill Harkins' directions. It was excellent. Experience is the best of all teachers.

By then I had become deeply interested in the art of home viniculture and realized that it was an important folk art.

When I went to Cooperstown, New York, to folklore meetings, I talked about this and soon I met a charming woman from Vermont who was also interested in the subject. She told me she had a few recipes handed down from her forebears, and she sent them to me. I began to inquire from my folklore friends throughout the country, and replies began coming in from every corner of the land.

Now I started to make wines from many fruits and flowers. Two other elements must be mentioned.

As a folklorist, I also became keenly interested not only in the lore of home wine making but also in the lore of the fruits and flowers that are the bases of these delightful drinks.

Here was a new realm, an unfolding luxury of lore and legend: how these wines came to be made, their religious and ceremonial uses, and the vast body of myths and tales entwining the fruits, flowers, and grains of which the wines are made.

When I was collecting folktales in Indiana and Illinois, I met a woman, the wife of a river-boat captain, who knew many tales of bygone days. Her home was filled with relics of boats: paddles, bells, clocks, mastheads, letters, a very hodgepodge of objects from along the Ohio River.

Throughout the evening she poured out tales connected with these many bits of wood and metal—remembrances of life on the Ohio. These objects were no longer just meaningless pieces of inanimate material. Each had a teeming life coming through the shadowy years of the river's history. Lifeless objects quickened into existence, thrusting their adventures into my mind.

The same can be said of the substances that go into wine making. Just as a knowledge of paints and their blending is a help and stimulation to the artist in his creative work, so a wide knowledge of fruits and flowers and their lore is a stimulating incentive in home viniculture. Of course, I have not put down here all the tales and all the lore that could be told of them; that would require several volumes. But I have tried to tell enough to add pleasure and knowledge to the making of your wine, and a little more joy to your creative ego.

To transform fruits and flowers and water (and a yeast) into an inspiring drink that lifts man above the earth is to enter the mystic realm of Dionysian divinity. There is an inexpressible glow on seeing the first foaming signs of the fermenting transformation. It represents an exciting new birth of life, bringing with it a world of multicolored feelings.

The very first thing I learned is that it follows the ever-common law of lore: recurrence. In every land of the world folks made, and are still making, wine from the fruits of trees and of the earth, from berries and from flowers. Hardly a farmer in any land of Europe, east or west, goes without his homemade wine in his cool cellar: wines made of grapes, pears, apples, blackberries, cherries, dates, palms, potatoes, wheat, and so on. In fact, anything that grows can be turned into a wine, and more wines are made in homes than are bottled by manufacturers.

You may ask: Why make wines from fruits and flowers when you can make wines from many varieties of grapes?

Men have said that grapes are the noblest source of wine. I don't dispute that a hair's breadth. The wines of France have brought symphonies of pleasure into my daily life. But because I enjoy the sight of blue-gleaming diamonds, or wine-red rubies or pea-green emeralds, would you hold against me the pleasure I find in Navajo turquoise? Every jewel has its realm of entrancing color, and every wine has its own exciting stimulation.

I, too, say grapes are the noblest source of wine, yet there is an adventurous richness in heavy-fruity elderberry wine, in the tantalizing bouquet of raspberry wine. A parsley or a peach or a seasoned tomato wine evokes new, fascinating gustatory experiences. You will have to take sip after sip to find a simile for the new tastes.

And, just as there is an enticing aura of dignified antiquity in the marching mythology of vine and wine, so is there a world of lore in fruits and flowers and the wines made from them.

The vine and its wine run in black and moon-silvered woods with the mystic rites of Dionysus and the bacchantes. In the red, living liquid there is the ancient symbolic chant of white-clad Hebrew priests and the offerings of brown-loined Egyptians with their quivering sacrifices.

But this great realm of poetry and literature of wine of the grape is equaled, if not exceeded, by the poetry, lore, romance, tales, and facts woven around fruits and flowers and herbs, sources of enchanting-tasting drinks. There are Silenus and sunshine, and nymphs and Venus, an endless roseate chain of romance and sentiment.

That is why I maintain that wines of fruits and flowers hold an honorable place beside the wine of the grape.

> "On the high Ku Su Tower the King of War was feasting.
> His slender queen Hsi-Shi, flushed with wine, dances,
> So fair and unresisting."

These lucid lines were written by the immortal Chinese poet, Li Po, in praise of great beauty, loving ladies, and fine rich wine—made of the fruit of the rice.

 "IN THE BEGINNING"

"IF HEAVEN LOVED not the wine,
A Wine Star would not in heaven course;
If earth loved not the wine,
The Wine Spring would not on the earth be.
Since heaven and earth love the wine,
Need a drinking mortal be ashamed?"
                                                LI PO

"In the Beginning"—thus the Great Book of the world commences, and thus should commence a book dealing with the divine drink of wine: a liquid that in a measure created a god, Dionysus, representing the life-giving light of the sky and life-giving moisture of the earth.

That the wine of the grape and the homespun wines from the

fruits of trees and garden are among the great forces of the world is proven again and again in the Holy Book. Great theological schools in the early ages argued this long and oft.

Note that the sun and the moon and the stars were brought forth by the Divine Will on the fourth day of creation, whereas herbs and trees and fruits were created on the third day. Proving, say I, that these were manifestly of greater importance to mankind than the constellations.

And if you ponder more deeply on the meaning of the wisdom of creation, you will agree with me that the Lord put Adam and Eve in the Garden of Eden for a divine purpose. He set them there to observe and study the flowers beyond their beauty and value for food.

"God planted a garden for him [man] to live in—wherein even in his innocency he was to labour and spend his time—which He stored with the best and choicest Herbs and Fruits the earth could produce, that he might have not only for necessities whereon to feed, but for pleasure also. . . ." Such was the belief of men nearer to the Bible by hundreds of years than we are. Observe the word "pleasure." Must not that signify that the good Lord, in His infinite wisdom, desired man and woman to learn the art of making wines from these flowers and thus partake of the inspired "delight" of life? I can see the four great archangels, Michael, Gabriel, Uriel, and Raphael, coming daily into "God's own plantation" to instruct our first parents in the art of viniculture from the things that grew in that first garden on earth!

So, from the beginning of the world, wines were not made from grapes alone. Far more were made from the fruits of the trees and the earth and from flowers in the green. The East made wine from dates; the Far East, from rice and palm leaves. In the lands of Europe, and in our land as well, it was fermented from fruits, honey, flowers, and berries. Thus it has been through the years. I have drunk apple wine in Austria, plum wine in Yugoslavia, pear wine

in France, and a strange yucca wine in Guatemala, and so on in all the countries in which I have traveled. Throughout the world, men and women take delight in creating wines for the pleasure and excitement and adventure they bring.

Adding to the interest of home viniculture are the many variations of perfumes and colors of flowers and fruits, which continue their aesthetic life in the wines. A rose-petal wine has the delicate scent of roses; and the same is true of rose-geranium and mint wines. Peach wine has the flavor and faint scent of peaches. This symphony of sight and taste, added to the warming exhilaration it gives, makes the "liquor divine" of which so many poets have sung.

Now, folks vary as the weather: each ethnic group, each locality, each individual has his or her own personal pet recipe, or the pet recipe of his or her great-great forebear.

Truly, wine making is a folk art. There is no more pleasant hobby-occupation for those who want to forget the tenseness of screaming business than to grow their own fruits, herbs, and flowers and to make wines of them. I know wise folks who have deliberately planted elderberry bushes along their fences and herbs and roses and marigolds and parsley in their gardens in order to make wine from them.

The Buddhists say: "Herbs, trees, and rock and stones enter into Nirvana." I am sure they are right. Probably that is the reason why I have taken to gardening.

There is also a very practical reason for making your own wines. A fine elderberry wine can stand sturdily alongside a good port or Malaga, and the cost of the elderberry wine is a fraction of the cost of the purchased vintages. The expenditure for the sugar and a little yeast covers practically all your outlay. The work is a pleasure. Before the advent of wines from France, and when they were very expensive, making wine in the home was as frequent as making relishes or dried apples or any home preserves. Strangely enough, these wines were made mostly by the women of the house

—during their "spare" time. While speaking of woman's work, I would like to say that now and then in this book I have allowed myself the luxury of a special cooking recipe, when it is related to the wine under discussion.

In conclusion, let me say that when you make your own wines you will join an enchanting universal brotherhood where all schisms, all differences of race, religion, and color are completely obliterated. It truly is a segment of the golden age in these troubled modern times.

The book is composed of three parts and a few cooking recipes, all interwoven. First, the art of wine making; second, the recipes; third, many of the facts, lore, and tales that twine around the fruits and flowers from which the wine is made. It is a modest attempt to present a rounded-out and pleasurable account of the complete art of folk wines from the very earliest times to today.

 AESTHETICS OF WINE MAKING

PERHAPS A TRAGEDY of modern life is the overemphasis on science, which sometimes seems to bring out the most brutish in man and blackens out the aesthetic and beautiful aspects of living. Since this is a book dealing with a graceful phase of existence, the aesthetic will have the foremost place. Therefore, before going into a few purely technical points, I want to say a few words about the winged part of folk wines.

First, let me mention the social and humane value wine making holds in the realm of occupational work or pastime or hobby.

There is no greater relief from the daily grind than the creative activity of making wine. It is both a quieting and a stimulating tonic for body and mind. And when a man is forced out of work into so-called retirement, at an age when he is richest in knowledge and experience, turning to viniculture will give him a well-merited and happy satisfaction. It discloses a large field for activity and adventure. Each wine is a new experience, for one can never tell what flavor will result from the work: Soil, water, climate, and seasons do not inform us of their influences.

If you have a home in the country, there is the pastoral pleasure of watching flowers grow and fruits ripen before you transform them into good wines. I spend many pleasurable moments following the growth of lacy elder blossoms, curling young vine leaves, scented roses, elephant-eared red rhubarb, and many other plants from which I make fresh wines.

There is the intimate tactile pleasure of gathering flowers, fruits, and leaves with your own hands, renewing the dim and veiled ancient feeling that there is a living spirit in each of them. You begin to know the value of rain and sunshine and clouds. All become close partners in your work, in which you are the creator.

One may also speak of the sensuous pleasure of washing the colored fruits in fresh, cold water and crushing them with the palms of the hands.

Just as Wordsworth said ". . . beauty born of murmuring sound shall pass into her face," so might I say the beauty born of intimacy with nature's art and marvels passes into our spirit.

Then follows the fascinating, mysterious fermentation, with bubbles ever rising in a rainbowy, springlike dance. And finally comes the ceremony of the first tasting and the warm feeling of success.

Thus, out of the fruits of the earth, out of the leaves of the trees, out of the wind and weather, you create a drink famous in history, famous in lore, and often famous in taste. You create a drink that gives inspiration for song, laughter, poetry, and great art. You have an undisputed opportunity to exercise your will in the choice of the medium, the color, taste, fragrance, bouquet. It is a hobby in which you are lord.

If your wine making is confined to a city apartment, you will have nearly all the same pleasures if you work with your mind and imagination even as you work with your hands.

Add to this a generous list of new gustatory experiences. I will never forget the sensation when I tasted for the first time the opulent, brown, Malaga-like elderberry wine. It made me think of wealthy olden Johnsonian merchants or Virginia gentlemen sitting around a table toasting lustily in loud voices. There is a Persian-perfumed feeling in rose-petal or rose-geranium wine. And with the first glass of parsley wine there was the surprising, exotic adventure of having come into a new world of taste.

Still this is not the end. Art, business, politics fade into air when you give your friends at the table the wine you have made. Invariably at first there is smiling skepticism. Then what a delightful change! A sip, a taste—an expression of shocked but pleasant astonishment, eyebrows raised in wonder that *you* could make so pleasing and potent a beverage. There is always a tiny tinge of envy that you have accomplished this. And for a long time it will be the only topic of conversation. There are requests for recipes, for details of the procedure.

When I bring to my table six or more bottles of different wines I have made, it is with a satisfying feeling of justifiable pride. And pride can be a fine virtue. The regal array of bottles of wine of my own making gives me a royal satisfaction, and so it will be with you when you bring out the wines you have made. It is a hobby that will pay in pleasures no money can buy.

 COMMON SENSE

I HAVE MENTIONED THE FACT that there are excellent practical inducements for making your wines at home. Show me the man who does not love a bargain at an auction or anywhere else; and making your own wine is the biggest bargain of any season. A little cost will fill many bottles.

The least expensive bottle of a rather poor port or an ordinary table wine costs $1.00 or $1.50. But for a much smaller amount you can have the finest dry rice wine or rhubarb or tomato or raspberry wine, or full, rich dandelion or elderberry wine, or a rose-bouquet wine of rose petals or rose-geranium leaves that is unbuyable at any price.

You can be a generous *seigneur* to friends, and the gift of a bottle of any of these wines is truly a princely gesture, for it elevates the spirit into shining realms and gives tastes of rare poetic and banqueting delight.

You must also be mindful of the good you do unto yourself and your friends when you drink these wines. Folk medicine teaches that carrot wine helps your sight; rhubarb wine is a fine sedative; dandelion wine will take away your headache from overwork—I have personal proof of that; and elderberry wine is a fine tonic. Nearly every fruit and flower wine is said to possess a medicinal virtue hallowed by custom and tested by usage from ancient days down to our own time. I wholeheartedly believe every one of these statements to be true. Wherefore I partake of every and all of these wines in generous quantities and with exemplary regularity, and thus escape all these ailments. This should be a good lesson and worthy example for any man or woman.

Follow carefully the recipes set down. None of these rules is complicated, nor are the necessary ingredients difficult to procure. All special yeasts are eliminated, for I have found that Fleischmann's bread yeast can be used for wines as well.

I have followed the same simple suggestions with regard to the utensils to be used, storage, etc.

My friends have held up to me the fact that it is very simple for me to make wines, because I have a home in the country. My answer has been, and is, that 90 percent of the wines made in the country can be made in a city apartment. One third of my own wines are made in my city quarters and stored in my bedroom, where the steam heat is never turned on. This winter I made wines in the city from oranges, grapefruit, carrots, pineapples, rice, almonds, celery, and figs, and all of them are excellent.

 WHEN

WHATSOEVER SEASON comes your way, there is always a lush, tempting fruit or flower from which to make a new wine. Even in the snowflaked winter days there are fruits appropriate for the purpose. Thus it is a year-round pleasure and pastime, ever changing, ever adventurous. Thus I have followed the "royal ordering" of the seasons, making wines to my own great satisfaction. So will you if you do likewise.

In the spring there are strawberries, and young vine leaves. There are dandelions in the lawn and along the road, crying to be taken up and transmuted into full, rich wine.

Then come lacy elder flowers and delicate cowslips in May, leading into glorious June, the month of roses and carnations.

Now fruits and berries begin to appear: raspberries, cherries, peaches. Reap that which is most abundant, or buy when the cost is lowest, and you will have wine at a minimum expense and maximum value.

Comes gleaming fall with its great harvest, and it is a great harvest for home vignerons. Then arrives soft winter with snowflakes dancing to the music of the winds singing through the fine fissures alongside your windows. (I always leave a tiny opening in my windows for the tale-telling wind music to enter.) There is the grand harvest of winter fruit for wine making. There are as many as during the summer, when vegetation is in lush growth.

At the moment of writing this, there are three crocks on a table in my kitchen, giving forth the lovely vinous aromas of carrots, oranges, and grapefruit, all dancing their bubbly dances.

I have just bottled my rice wine, clear and golden. I will make my next from caraway seeds and then from dried apricots and other preserved fruits, until the warm seasons bring their new crops.

WHAT YOU NEED

Crocks

THERE ARE JUST A FEW simple utensils needed to become a successful viniculturist in folk wines.

First, the vessels in which you actually ferment the wine. There are two kinds of containers you can use. One is wooden barrels. They can be bought in almost any size and are very expensive. I used them when I began to make wines, when they were inexpensive. If you have the space and don't mind the cost and the extra work that must be put into caring for them, why, bless you, go ahead and use them. I have completely discarded them and use old-fashioned crocks instead. They have just as much romance and beauty as scented wooden barrels.

I could write a book on crocks in general and some crocks in particular: old ones, gray-brown in color, with honorable stains of usage and history. There is a homespun artistic beauty in those old vessels used in bygone days for dill pickles, tomatoes, and relishes. They are brilliantly glazed and smooth and pleasant to the touch. The blue pictures of strange birds and the fearlessly Japanese incomplete brushed leaves and designs possess a wildness of artistic adventure that captures the spirit of folks of those times.

The names of makers and cities are openly and fearlessly displayed. W. Roberts, Binghamton, New York; J. & E. Norton, Bennington, Vermont; N. A. White & Son, Utica, New York; and many more. All sturdy names, sturdy workmanship, with enough of the artisan's interest to make each individually distinctive. The sheer universal color of blue, always used, a color of luck and good hope common to mankind all over the world, is folkloristically significant. I saw that same color in peasant homes in Yugoslavia, Switzerland, Turkey, India, almost everywhere, including our own land. The use of these particular vessels for wine making gives an additional zest to your delightful avocation.

Get crocks in two- to six-gallon sizes. Larger ones are too heavy to handle. The number of gallons a crock holds is always marked on the outside.

I use four- or five- or six-gallon containers even to make only two gallons of wine. Keep in mind that the bulk of your fruit and the sugar require additional space. The cost of these crocks runs from $3 to $5, depending on the condition and the excellence of the decoration. I have bought mine at antique dealers while traveling up and down the New England and Middle Atlantic states collecting folktales. Try to get covers for the crocks, even though they do not match. These are very useful.

If you can't find old crocks, you can buy modern ones in department and hardware stores. They are just as serviceable, even though there is nothing romantic about them. You can also use unchipped enamelware. Enamel is more expensive and chips easily. I use it only for the boiling I have to do.

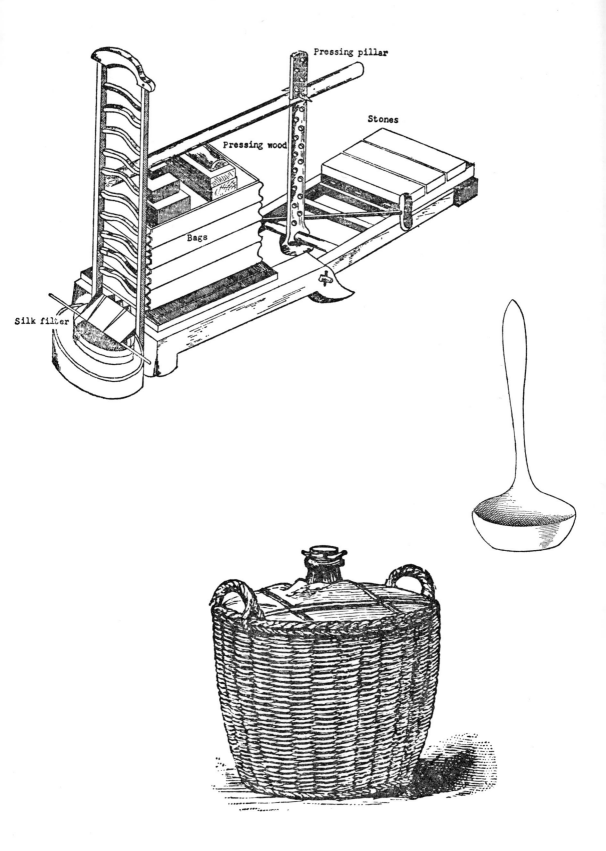

Never use metal containers. Never use metal anywhere in your wine making. Fermentation in metal will give the liquid a most unpalatable taste and might even make it poisonous. Even aluminum discolors water, and since color is an important asset of wine, anything that destroys color should be avoided.

Glass Jars

These are important. You need two to four half-gallon or gallon glass jars with wide mouths and covers. Some hardware stores have them. If you cannot find them in stores, go into the kitchen of any restaurant you frequent and ask someone to give you a few empty gallon mayonnaise or pickle jars with their lids. This is the way I get them, paying only a small gratuity. Such jars are of great value for straining the wine.

Funnels

You should have three funnels of various sizes. Plastic are the best; they are the lightest, are easy to store, and have the divine virtue of not breaking. But, of course, you can use glass or enamel if you prefer. Do *not* use metal.

Spoons

One or two large wooden spoons, which can be bought in any hardware store, and two plastic spoons are of value for mashing and for removing scum during the fermentation period. Glass or unchipped enamel can also be used, but those are more expensive and break or chip easily. Again, do not use metal spoons.

Straining Cloths or Bags

Good straining bags, holding about two quarts of liquid, can be made or bought. You can make your own from unbleached heavy muslin or muslin-flannel. Clean, good kitchen towels can be used for the same purpose.

For coarse straining, use a sieve of any material except metal.

Scales

It is good to have a scale that will weigh from ounces to pounds, but it is not necessary. Sugar can be weighed on a bathroom scale.

Press

A press is not necessary unless you go into large-scale wine making. You can make up to ten gallons of wine without a press, even when you use apples, grapes, or rhubarb—the three fruits that need strong pressing. The use of your hands is good and pleasant exercise, and since the vintner of homemade wines rarely deals in large quantities, this is a simple task. Tactile sensations and the pleasures they bring are rarely mentioned except among the blind, yet they are important in our lives and have an important place in the folklore of the world. When you feel and see the soft pulp of fruit in the palm of your hand, when you see the perfumed juice of these fruits running through your fingers in lovely tinted colors, it brings a triad of Dionysian delights, visual, tactile, and olfactory, heralding the coming good WINE. You can finish the pressing by squeezing the fruit through a cloth.

If you happen to have a crock with a spigot at the bottom, you can put the fruit in it and press it down with a piece of wood that just fits the inside of the crock, the liquid running out through the opening at the bottom.

There is an excellent small hand press made in Switzerland that can be bought in a good hardware or department store. There are stores in New York City and London that specialize in materials for wine making. Probably most large cities with an Italian population also have them. There you can buy all things needed for your home winery.

Hydrometer

This is another instrument that you really do not need but which you can buy if you are chemically-minded or if you make wine on a large scale. The hydrometer tells the sugar content of the wine. It can be bought in any good chemical supply store or in the shops I spoke of before. I have one but I rarely use it, relying on my taste and experience to tell me how much sugar I need.

Thus, it narrows down to: (1) two crocks; (2) glass jars; (3) some wooden and plastic spoons; (4) some funnels; (5) bottles and corks; (6) enamel pots for boiling; and (7) materials or bags for straining. None of these is expensive, and all can be procured easily.

To Help You

A list of places where you can get advice or materials is given on pages 399-403.

THE LIVING FRUITS AND FLOWERS

Recipes

NOW THAT YOU HAVE surrounded yourself with faithful, long-living utensils for your artistic avocation of viniculture, you still need a few helpful suggestions before you are ready to begin the work.

The recipes in the book are arranged within broad categories, alphabetically. There are no recipes using fruits too difficult to obtain. Since your time and space are often limited, the planning of the work is important.

Of course, variations are possible, but you should try those only after you have had some experience.

It is a good idea to copy the recipe on a slip of paper and then check each step as you proceed. Be sure to include the date you begin and the date for straining. Make notes of additions, changes, or anything unusual that comes up. This will be of help later. Fasten the slip with Scotch tape to the crock cover so that you can refer to it easily.

Follow the recipes carefully at the beginning; after you have had experience, you can play with variations to suit your taste. For example, when you make orange wine, you can vary the thickness of the rind so as to add a piquant, slightly bitter flavor. Similar variations are possible with other wines. All of this will come with time and experience.

Quality and Treatment

Choosing the fruits, vegetables, and flowers, whether you grow them or buy them, is of great importance, for on their goodness will depend the success or nonsuccess of your wine. If they are bought, they must be fresh and ripe but not overripe, unless the recipe calls for it. Strawberries, pineapples, cherries, celery, carrots—all must be in perfect condition. Cut away any parts that are not healthy. Even a small amount of decay will spoil the wine.

If you grow your own flowers or fruit, the same degree of care must be exercised. Flowers, fruits, leaves, and plants must be picked at the proper time, and when they are dry. Rain and dew destroy the essential oils; this will rob the wine of much of its richness.

You must rinse and wash the fruits and flowers unless they were never sprayed and do not grow near dusty roads. Everything washed must be dried. Wet fruit or leaves may cause the growth of mold, which will guarantee spoilage of the wine. Some flowers and herbs must be crushed first; this will be indicated in the recipes. There are instruments with which flowers and fruits can be crushed, but I think it is much more pleasurable to do it by hand.

Quantity

The amount of fruit or flowers you need is stated in the recipe. Of course, this depends on the amount of wine to be made.

Most recipes here are for two gallons of wine. Chances of failure are so small that one may safely begin with that amount unless one likes to experiment with less, varying the quantities proportionately.

Remember that the juices of the fruits add an appreciable amount of liquid; sugar also adds bulk. Cereals, on the other hand, absorb a certain quantity of liquid. These variations, however, will not interfere with the process of fermentation. What must be kept in mind and allowed for is the bulk of fruits and flowers, such as peaches, tomatoes, berries, marigolds, etc. Therefore, to make two gallons of wine a vessel holding four gallons should be used.

You won't get more than eight or nine bottles of wine from the two gallons you set out to make. The loss comes from the skimming during fermentation and from clearing and decanting afterward. This will still yield a goodly amount for personal use and to put away for long maturing. If you want enough for gifts, you must make more.

Increasing the Amount of Wine

You can increase the amount of wine in a very simple way without diminishing its excellence. Here is how it is done.

After squeezing the juice out of any fruit, berry, or flower (except cereals such as barley or rice or potatoes), take the pulp and return it to the enamel pot. Now add half a gallon, or a little more, of water and put it on the fire. Let it simmer for about half an hour. Taste it, and when the liquid has the flavor and bouquet of the fruit or flower or herb, strain it into the must (as the liquid that will turn into wine is called). Of course, this must be done before the yeast is added.

I do this very rarely. Why? I like my fruit or flower wines undiluted.

 THE MYSTERY OF WINE

The Yeast

WINE IS MADE BY yeast in combination with several substances, but many of the fundamentals of the process are still a mystery.

Yeast is the name given to a living organism. It lives as a single-celled creature so small that it can be seen only under a microscope. Any one cell grows until it reaches its characteristic size, and then it bulges out to make a little bud at some spot. A yeast cell with a bud looks like a large and a small soap bubble incompletely separated. The bud drops off; each is then a completely new yeast cell. Any one yeast cell produces a large number of buds and therefore creates a large number of yeast cells. In the course of growth, astronomically vast numbers of cells can be produced in small volumes of space. It would not be at all unusual, for example, for a thousand billion yeast cells to be in a glassful of the fermenting liquor—the intermediary stage in making wine. This multiplication of yeast cells is an important part of the fermentation process that creates the wine.

If one started out with a completely sterile fruit, made sterile juice of it, then added one yeast cell, it would be sufficient, if given plenty of time, to produce all the billions of cells needed for good fermentation.

It is not really necessary to add yeast to a fruit or berry to produce fermentation; they already have a large number of yeast cells and various other types of cells on their skins, which would produce biological action. Then why add yeast? The addition is necessary in wine making to get a head start in the competition with the various bacteria that are also present, and which battle to have lives of their own.

The same mixture that supports the growth of yeast is also a fine medium to support the growth of other bacteria that would be very harmful for your purpose. You would hardly find pleasure in drinking a liquid resulting from a bacterial fermentation other than yeast, and tasting—shall we say—like Limburger cheese, which is also the result of a fermentation process. By adding a large number of yeast cells at the beginning, the yeast population in the liquid will grow so rapidly and so vigorously that no other microorganism will have a chance to compete to any serious degree.

In a mixture of fruit juices or fruits or flowers where the fermentation process takes place, two things happen. First, the yeast cells begin growing and dividing. Second, in order to gain energy to grow, they consume the sugar present or added to the liquid.

Here a fascinating activity enters, where the seemingly expressionless cells show "reasoning" or, may I say, knowledgeable selectivity.

There are two ways in which sugar is used by yeast cells. If air is present, they burn or oxidize the sugar completely, forming carbon dioxide and water. This is what happens to sugar in our bodies. But if air is not present, the sugar is burned only incompletely, forming carbon dioxide, and the rest of it turns into alcohol. (Pasteur defined fermentation as *"la conséquence de la vie sans l'air."*) And so air must be excluded in order to create alcohol.

However, we don't actually take away the air from the fermenting liquid—we let the yeast do that. At first there is a great deal of air dissolved in the fruit juice. But as the yeast cells grow, they use up this air rapidly. All that is necessary now is to keep the fermentation liquid still, and the bulk of the liquid will have virtually no air in it. A small amount of oxygen penetrates very slowly from the very top, which is exposed to air, but this is used up rapidly by the yeast cells growing there. In the absence of stirring, almost all the liquid lacks air. For that reason it is very important not to disturb the vessel in which the must is fermenting.

So, living and growing, the must is fermenting. The hosts of yeast cells are busy at work converting sugar into alcohol and other molecules of odors and flavors that make of wine a creation of art and pleasure. In the absence of air, the alcohol content increases in amount and strength. Finally it reaches a concentration so high that even the yeast cells cannot tolerate it. And so, alas, the yeast cells are destroyed by the power they create.

The concentration of alcohol that will destroy yeast cells is roughly 7 to 14 percent. Wines with a higher alcoholic content have been "fortified" with extra alcohol over and above what was created by their own yeast populations.

Interestingly enough, the study of the fermentation process has been both the beginning and the cornerstone of modern biochemistry. If you want further, more detailed information about yeast, you can find it in textbooks on biochemistry or physiology or in other books and publications dealing with biology.

Sometimes other nutrients, such as tartaric acid or tannin, are added to your mixture, but this is not really necessary. Fruits and flowers have enough of their own proper nutrients for fermentation.

There are many kinds of yeast that can be used. In England,

special yeasts for wines have been richly developed. They are sold by British firms specializing in wine accessories. I am listing a few of them at the end of the book. Here in America, the Berkeley Yeast Laboratory also manufactures special yeasts for wines, and Mr. J. H. Fessler, the director, most generously gave me a good deal of information about its products.

Professor G. Levine, of Cornell University, kindly sent me some wine yeasts he prepared especially for me. And finally, my son, who made the first wine for us, gave me some specially prepared yeasts as a holiday gift.

I also procured specially prepared yeasts from England. I tried all of them. Before, I had made my wines with regular Fleischmann's active dry yeast used for baking, with very good results. Next I began working with the special yeasts, both American and British. It took time and care, but I did not mind this in the least, for I enjoy experimenting. Then I learned that in my own vinicultural efforts, it made no difference which yeast I used. So I decided to return to the ordinary American yeast since it was the simplest, the least costly, required the least amount of work, and gave satisfactory results.

Each package of Fleischmann's yeast holds one-quarter ounce, which is equivalent to a level tablespoonful. It is important to note the date of expiration, which is always printed on the package. Do not use the yeast after the indicated date. Remember that yeast is a living organism and will not live forever without food and water.

There was one nutrient I procured from England that helped when fermentation proved a little slow. This came in the form of tablets called La Claire Wine Yeast Nutrient.

Using the Yeast

Each recipe tells the amount of yeast to be used. I always refer to the Fleischmann product described above. Generally, the amount called for will be ½ or ¾ ounce.

CHERRY

Now, yeast must be prepared properly. I use ¼ ounce to 1 gallon of liquid. Pour the yeast into ½ to ¾ cup of warm water and stir it with a wooden or plastic spoon until it is completely dissolved. Do not use hot water; that kills the yeast. Do not use cold water; that retards its activity. You can, if you wish, dissolve it in lukewarm must. I generally use warm water.

If the recipe calls for boiling the liquid, let it cool until you can put your finger in it comfortably before adding the yeast.

If there is no boiling and the liquid is cold, pour half of it into an enamel pot and heat it, then pour it back into the cold liquid; this gives the right temperature. Pour the dissolved yeast over the lukewarm must.

Many home wine makers first prepare a piece of toast and pour the dissolved yeast on both sides of it, then put it into the must, pouring the rest of the yeast all around. This is said to help the fermentation. I have made over half my wines without toast, and there seems to be no difference in the result.

In the recipes I sometimes suggest the use of toast and at other times I do not. This simply means that I used toast when I made those particular wines. But I feel that with the proper sugar content, either in the fruit or added, and the proper temperature during the fermentation period, the toast can be eliminated. Practice and experience will teach you, as it has taught me.

If, after a day or two, the liquid has not begun to froth and bubble, I often crush a tablet of La Claire Yeast Nutrient and put it in.

The Vinous Fermentation

"This may be said to be a DIVINE operation which the Omniscient Creator has placed in our cup of life, to transmute the fruits of the Earth into wine, for the benefit and comfort of his Creatures." So said P. P. Carnell, Esq., in 1814 in his *Treatise on Family Wine Making,* and so say I today.

Wines from fruits and leaves and flowers made in all the lands of the world have been part of ever-moving social history and of all people's lives. Festivals, songs, dances, and ceremonies have been part of wine making and wine drinking, thus enlarging the horizon of pleasurable existence. In all religions, the ideal life has been located in gardens filled with fruits and flowers: the Garden of Eden; the Garden of the Blessed. And from these gardens we derive the omnipotent blessing of wine.

According to my son, upon whose knowledge of biochemistry I depended so heavily for a description of yeast chemistry, the mysterious process of vinous fermentation can be loosely broken up into two divisions: first, the major components; and second, the trace components.

The first are the original materials: fruits, flowers, vegetables, herbs, alcohol, sugar, and water. These are easily measured and their sources are well known.

Then comes the second group, the trace components, which create the wine as we know it. This is an area of great complexity and mystery, even in the present advanced stage of biochemistry. The total number of these components is not known; the chemical structures of many of them are unknown; and the exact pathway for the formation of many of them is unknown. Even of those that are known, the exact proportion of them in wine is unknown.

These all-important trace chemicals create the tastes and the aromas that enhance the pleasures of drinking wine. Just yeast, alcohol, and sugar could never do that.

Some of them are probably present in the original major components. That is why each flower or fruit gives a distinctive flavor, and perhaps aroma, to its wine.

Practically every fruit and flower contains its own yeast and sugar for fermentation. But, as I stated before, they do not contain enough of these elements to develop the necessary alcoholic content to make wine. That is why we must add both.

Sugar is added usually at the very beginning of the process, and sometimes as the action goes on. Each recipe tells how much is to be added and when. Grapes have so much yeast and sugar of their own that neither has to be added.

The actual time of the visible—what I like to call "boisterous"—fermentation differs for many reasons. Temperature, type of ingredients, yeast, and the seasonal element all contribute to this variation. The process generally lasts from seven days to four weeks.

Some home wine makers suggest straining the wine when in full fermentation, that is, four or five days after it has begun to work, and letting it finish clear. I don't do this. I let the visible fermentation continue with all the ingredients in the liquid (must). When there is no more visible action, I do my straining.

But the straining does not end the fermentation. Invisible fermentation ("maturing" is a better word), which makes wine so appealing to the palate and the imagination, goes on for days and weeks, and I believe continues for months and years in the tightly closed bottles. There is an ever-changing process of created vigor, flavor, and bouquet. Taste a wine one year old and taste one three to five years old, and note the change in sharpness and sheer brightness between the older and younger wine. With the years, wine becomes lively as well as lovely.

If, when visible fermentation is ended, you find the wine too dry, you can easily sweeten it by adding honey or sugar syrup, which is very easy to make.

At this point, some viniculturists suggest adding a pint of brandy for each gallon of wine. I think this wholly unnecessary. Each wine can stand on its own.

For the success of the fermentation process, you must keep the liquid at the proper temperature. If it is too cold, fermentation will stop. The temperature must not be higher than 75° and not lower than 60°. Above all, the liquid must not be in a place where it gets cold at night.

The amount of sugar necessary needs consideration. You can vary the amount to suit your taste. A high sugar content will result in a heavy wine; less sugar makes it drier. For each gallon of water, use 2½ to 4 pounds of sugar. You will learn with experience how much sugar you want. A good deal depends on the fruit used. Thus, I used 4 pounds of sugar for each gallon in making my tomato wine, and the result was a very dry wine, whereas the same amount of sugar with elderberries produced a very sweet wine, so I now put in only 2½ pounds of sugar for each gallon of the elderberry. The same holds true of many other fruits.

White granulated sugar is the most useful. Sometimes I use brown sugar, say with elderberry wine, for this adds a brownish, pleasant taste and color. Each recipe will tell you what kind and approximately the amount of sugar to be used.

The recipes also tell the approximate amount of time for fermentation. Sometimes this is over in a few days, sometimes it takes longer. It is best to let the process take its own time, but when it is overlong, all you need to do is to put the wine in a cold place, or add a small amount of liquor with a high alcoholic content, and it will stop. About half a glass to a full glass of liquor per gallon of wine is enough.

Sugar Syrup (Sometimes Called Gome)

The addition of a dilute sugar syrup is the simplest way to sweeten wine or liqueurs. (Of course, you can use granulated sugar, but this takes a longer time and more work.) It is best to keep such a syrup on hand if you make many wines and liqueurs. Or you can prepare it fresh every time you need it. Sugar syrup is made by dissolving and boiling a large amount of sugar in as small an amount of water as possible.

Take, say, four pounds of white granulated sugar, put it in two pints of water, bring it to a boil, then let it simmer for at least twenty minutes. This will make an excellent thick syrup that can be added to any wine for sweetening to taste. You can also dissolve two pounds of sugar in a little less than two cups of water. Some folks add the white of an egg, well beaten. Impurities of any kind, at any time, should always be removed.

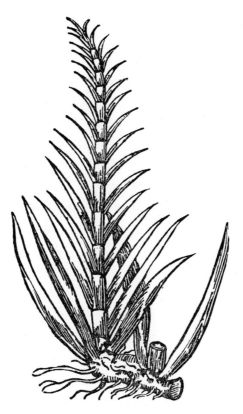

SUGAR

 HIGH ADVENTURE

Sweet and Dry

YOU HAVE HAD A glimpse of wine chemistry to give you an inkling of what you are doing; you have decided on the wine you want to make and the quantity; and now you must make the broad decision whether you want it dry or sweet. This is regulated largely by the amount of sugar added. Each recipe will give the amount to be used for a good vinous taste. But a word of warning: Every now and then, even when using the right amount of sugar and following all the laws that make the wine you set out to make, something mysterious happens and the result is entirely different.

The only explanation I have given to myself is folkloristic and I honestly feel it is as good as any. It is all caused by the "Wine Mother." It is she who decides whether the wine is to be dry or sweet, and she often determines other characteristics as well, such as strength and taste and flavor. I said before that there is an unusual element of adventure in wine making. You enter into an unexplored gustatory realm to some extent, uncontrolled by mechanical and scientific precisions and measurements.

Countless people throughout the world believed (and many still believe today) that every useful plant and tree is animated by a divine being often called "mother," which is its life and growth. And so wines made from fruits and flowers must perforce have this divine "mother" that directs their varied lives. Let me give just one first-hand example of one of my vinicultural adventures.

I made a carrot wine one winter that turned out to be a fine, golden-yellow, sweet dessert drink. I liked it so well that I decided to make two more gallons to give to friends. I used the same amount of sugar, yeast, raisins, etc., and the result was a dry, sharp, pleasant wine approximating a Graves.

Of course, there is the possibility that the carrots I used for the first wine may have been very much sweeter than those I bought the second time. They may have come from a different locality and may have grown in a different climate.

Such distinct variation is usually not the rule. Sweetness and dryness can be controlled by the amount of sugar used. Also, experience will help you to judge how much the amounts can vary.

Raisins

The addition of raisins is an excellent help in the fermentation and flavoring of wine. They should be cut up as finely as possible so that the meat of the fruit has an opportunity to do the requisite work.

But the raisins we buy here are difficult to chop or cut finely. I have tried cutting them with heavy and light scissors and chopping them. None of the methods gave a perfect result. Even so, they served well the final purpose for which they were intended. So I always cut my raisins the best I can, and you will do the same.

Taste and Flavor

Here you enter a vast field of delicate gustatory nuances that are a delight to explore. Fifty percent of the result will be in the lap of Happy Accident or the Mother Spirit of whatever fruit or flower you use, and 50 percent will be your own work. It is interesting to note that homemade fruit wines have a much better taste and stronger "personality" than the same kind bought commercially.

First, the elements—rain, sun, and even soil—and locale will condition fruits and flowers to variations in taste. There is a great variety of subtle mutability in this.

Using different recipes, many fruits and flowers will yield wines having interesting variations in flavor and taste. Thus, you can make different rhubarb wines, different raspberry wines, and so on.

You can also mix different berries, fruits, and flowers and thus produce different exotic shadings in your wines. Gooseberries and currants will create a piquant, delicate-flavored drink; the addition of apples will bring sparkling changes. Pineapple-sage leaves added to parsley bring a kingly richness to the wine.

It is an excellent idea to taste the forming wine now and then during the fermentation period; then you can change the flavor according to your taste. You might want to sweeten the wine or to add additional fruits or flowers to heighten a particular flavor. Orange wine might need more oranges, or mint wine might need more mint to strengthen the cooling green taste.

The addition of fresh ginger root—not too much; an ounce to a gallon—or a few cloves will impart an intriguing, spicy Eastern flavor to the wine. These can also be added after the wine is bottled, if you like a spicy wine.

In Scotland, where home wine making is very common, the addition of spices to wines is favored. Miss Florence McNeill's excellent book, *The Scots Cellar,* will give you a great many Scottish drinking customs, Scottish drinks, poetry, tales, and recipes. It is a delightful as well as an instructive volume.

Another spice that will give the wine a delightful zip is a few peppercorns, say four or five added to each bottle. They also help give a greater clearness and dryness. Mustard seeds have the same function.

Sometimes you may find your wine a little sour because of an overamount of acidity. How did it happen? Don't rack your brain too much about it. Endless things happen in life for causes we never know. Besides, this can be corrected. Put a spoonful of barley in each bottle and let it stand in a warmish place (60°-70°) for about two weeks. Then strain it.

Or you can put a teaspoonful of raisins in each bottle and let it stand for about four weeks. If the wine is in gallon jars or bottles, put in half a pound to a pound of raisins. This will also eliminate flatness. Raisins will color the wine a little. After a month, strain the wine, and the sourness will be gone. If the wine is oversweet, add to it some rice wine and let it rest for a time—say four to eight weeks before bottling. Or you can put 2 or 3 ounces of brown rice into each quart bottle, let the wine rest for two or three weeks, and then strain it.

If you add wheat or potatoes to any wine during the fermentation period, it will add greatly to its potency.

Should you like a sweet wine, it is simple to add sweetening *after* it is fermented. The easiest way is to add sugar syrup or gome, as described in the previous chapter.

If your wine turns to vinegar, as happened twice to me, really the wisest thing to do is to use it as vinegar. It is much better than any vinegar you can buy.

Why did it turn into vinegar? There are many reasons. It may have been exposed too much to the air. Perhaps there was not enough sugar. The fermentation may have been too slow. It is difficult to tell the exact cause.

You might try to correct it, but it is hardly worth the time. A pinch of slaked lime is said to be helpful, or adding 1 ½ grams of potassium bitartrate to each gallon, three or four times at three-day intervals.

There are some specially prepared remedies for turning vinegar back to wine. These can be bought in stores that sell wine-making supplies. But I don't think it is ever truly worthwhile to try to make a wine out of a vinegar. I say again—it will do much better as vinegar.

How Strong?

The strength of your wine is controlled. The average homemade wine ranges from 7 to 14 percent in alcoholic content. Those made with cereals and potatoes are a little higher.

The alcoholic strength can be increased by the addition of brandy or alcohol. Many authorities on the subject suggest the addition of a little brandy to almost all wines made at home. This probably began in England with the idea of imitating imported wines when they were very expensive and when ports, sherries, and other wines from foreign lands were very "distinctive."

A "fortified" wine is simply one in which the alcoholic strength is raised. These are not table wines to accompany good eating.

Let me say again, and finally, that I have never found it necessary to add brandy to any wine I have made. If you like to do so, add about one half to one glassful of brandy for each gallon of wine.

Citrus Fruits

The rind and juice of lemons or oranges, or both, is added at the beginning to almost all wines, as the recipes indicate. There are two reasons for this.

Yeast needs acid to make it function well. Many fruits and flowers do not have enough of this acid, hence the addition of lemon and orange rinds and juice, which gives the needful acidity. The second value of citrus fruit is that it adds a tangy flavor to the wine, a sort of zip that cannot be described easily. Suffice it to say that it contributes to the pleasure of the taste and the drinking. Even adding a little lemon juice, not much—a teaspoonful to a bottle—perks up the flavor.

The one important thing is to cut the rind as thin as possible. The white pith under it will cause a slight bitterness. Those who like that bitter touch—I am one of them—can cut the rind thicker now and then, and so vary the flavor.

Use only lemons and oranges that are not overripe.

Just as you can color grape wines, so you can vary the color of fruit and flower wines.

The addition of beetroots or raspberries will give you a rosy wine.

If you want a brownish color, add tea. Spinach leaves will produce a delicate green. Marigolds will create a saffrony golden red.

Boiling water usually changes colors, whereas cold water preserves colors as well as flavors and scents. Use boiling water only when directed. I made a tomato wine using boiling water and it came out almost a clear white, yet boiling red apples preserves their color.

Heavy and Light Wines

The consistency of wine can be regulated. You can make a wine light, medium, or heavy by the amount of fruit or petals or cereal you use.

For a *light* wine, use 2 to 3 pounds of fruit or berries per gallon of water. Two quarts of petals per gallon of water will suffice.

For a *medium* wine, use 3 ½ to 5 pounds of fruit and 2 ½ to 3 ½ quarts of leaves to the gallon.

For a *heavy* wine, use 6 to 8 pounds of fruit, and 4 to 6 quarts of leaves to the gallon.

When you make a wine for the first time, I suggest that you follow the directions I have given. After that you can add or subtract or vary to your heart's content—the only limitation will be your own imagination.

 THE WORSHIP OF BEAUTY

Straining and Fining

BEAUTY IS AS NECESSARY to daily life as is the shining sun, and we are all sun worshipers. The beauty of clear raspberry-colored wine, the pale amber of a rice wine, the spring-green of a mint wine enrich the pleasure of the drinking a hundredfold.

When wine is the taste and color desired, the next step is to clear it of sediment, debris, and cloudiness. A cloudy wine is neither pleasant to look at nor pleasant to drink, wherefore you must take the time, the patience, and the effort to make it gem-clear. For your comfort and mine, here is a plan that saves a good deal of time.

Begin with general straining to get rid of the visible debris. Experience has taught me a simple method that will help you do this with expedience and dispatch.

I do my first straining from the crock into gallon or half-gallon glass jars with wide mouths, or into regular white-wine gallons.

First, remove with a wooden spoon the toast and any other material that may be on the top. Then strain the liquid through a funnel, if you are putting it into glass gallons. Put either absorbent cotton or a piece of muslin into the funnel. Straining can also be done through filter paper, but this I reserve for the final straining. I use a large funnel and an enamel dipper or a porcelain cup to pour in the wine.

If you use glass jars, as I suggested, put a strong piece of muslin or a hand towel over the wide mouth of the jar, curved inward a little. A funnel and filter paper can also be used.

This takes time, and in order to expedite the work, I generally have two or three jars or gallons that I use simultaneously. If time hangs heavy while waiting, read some poetry, preferably Li Po.

There is still a third way of doing this first straining. Make, or buy in a wine-equipment store, a three-cornered bag holding a quart or a little more. Fill it with the wine, tie it with a cord, and hang it from a chair over an enamel vessel or a crock, and let the wine drip through. Do not try to hasten the straining by violent squeezing of the bag. If you are impatient, just press it gently. The amount of liquid that remains in the bag is negligible.

Don't try to get the dregs out of the crock. It is better to lose a little wine than to have a cloudy wine with a harsh taste.

Using big glass jars or glass gallons enables you to see how clear the wine is and to note if there is any fermentation going on.

Cover the jars or gallons and let them rest in a coolish place—a cold temperature helps to clear the contents. Let the wine stand thus from two to six weeks. During that time, whatever fermentation is necessary continues, and all the sediment sinks to the bottom. If the cloudiness persists, it may be caused by the fact that the alcoholic content of the wine is low. In that case, add a little brandy, 4 to 8 ounces to a gallon, or any beverage of high alcoholic content that has no distinctive taste, such as vodka. This will generally expel the cloudiness.

But more often such murkiness is caused by particles of the fruit used. Then it needs fining, which means clearing the wine.

But before going into this phase—suppose your fresh wine is clear except for sediment at the bottom. Then siphon off the clear liquid into bottles, using a glass or rubber tube, and fine the rest. This rule should be observed with every wine.

There are many fining agents, most of which can be bought, already prepared, from the English firms previously listed.

Let me say here that if you like a clear, crystalline wine, it is often advisable to fine the part you have siphoned off, as well, for when you let this part rest—and you should let it rest from three to five weeks—you will find a fair amount of sediment at the bottom of this, too.

I have used bought fining agents and they are excellent. But what is more simple and certainly less expensive is common eggshells or isinglass.

The use of eggshells is the simplest method of all (but do not use them in rhubarb wine). Take the shells of freshly opened eggs, with a little of the white still in them, and put them into the wine. Let this rest from five days to two weeks, and you will see that the shells are covered with dregs and the wine is clearer.

Here is a suggestion for siphoning that will prevent any sediment coming with the liquid. Get a short plastic knitting needle or a thin, clean piece of wood about one-quarter inch or less in diameter, and attach it by a rubber band to the end of the rubber tubing, allowing it to extend one-half inch beyond the tubing. Set the wood on the bottom of the glass jar or gallon, and the end of the rubber tube will be well away from the sediment. Remember again: It is better to sacrifice a little wine at the bottom of the jar than to have it beclouded.

Isinglass is perhaps the best fining agent of all, and it is inexpen-

sive and very simple to use. It is the purest form of gelatine, made from fish bladders, and dissolves readily in water. The most important warning is to use it sparingly. If you use too much, it will make the wine even more cloudy.

You can get isinglass either in powder form or in sheets at any good pharmacy. A quarter of a pound will last for years. For a gallon of wine, dissolve about one-half teaspoonful in a little hot water. Pour it into the wine and stir gently with a wooden or plastic spoon. The isinglass will then sink very slowly to the bottom, carrying with it the suspended materials that cause the clouding. Let the wine stand quietly for three days to a week, and then siphon off the clear part.

If, after the wine is in the bottles and has rested, there is still a sediment, siphon off the clear part—about two thirds of the bottle—and filter the rest through filter paper. This will give a perfectly clear wine.

Put the filter paper in the funnel (glass or plastic, not metal) and set it into a quart jar. Fill the funnel with wine and turn to some other work, for it will take many hours to seep through. Finally there will be a brilliant clear wine in the jar and a brownish sediment on the paper. This wine then should be poured into a bottle.

There is always a small loss of wine in the process, and since it is important to have a bottle filled to within one inch of the top, I always add a few spoonfuls of white Bordeaux or, preferably, dry sherry. This amount will not alter the original wine.

Let me summarize this very important part:

1. After the visible fermentation has stopped, strain the wine through any good clean muslin or a hand towel into half-gallon or gallon jars with wide mouths, or into clear wine gallons. Fill them to the very top and let them stand at room temperature for two to four weeks, removing any impurities that may rise to the top. If you remove an appreciable amount, top the wine with some extra wine you have, or with dry white wine or dry sherry. As I said, the ad-

dition of such small amounts of wine does not alter the flavor of the original wine.
2. After two weeks, siphon off the wine that is perfectly clear into bottles; fill them to within one inch from the top, and cork them by hand.
3. Let the filled and corked bottles rest for a week. If, at the end of that time, there is sediment at the bottom, siphon off the clear wine and run the rest through filter paper. Fill the bottles and close.
4. If, after the wine has rested for three days, there is still some sediment at the bottom, rack it as explained in the next section. This racking can be repeated, if necessary, four to six months later.

Racking

Racking is the simple process of pouring off the clear wine, free of sediment and impurities. It is done most easily by siphoning or by taking off the clear part with a dipper. The first method is the best. It will assure you a drink free of yeast particles and other kinds of fruit debris that would interfere with the good taste of the wine.

It is wisest to rack wine several times, allowing long intervals between. The ideal way is to let the wine rest three months for a final racking. Such removal of the lees also helps the wine mature. Often I find it necessary to rack the wine from a bottle I bring up from the cellar. When you rack wine you must always lose a little of it, and for that reason it is advisable to leave one bottle to fill up that loss. This is called "topping." You can also use for topping a neutral-tasting wine of your own making, like rice wine, or a purchased dry white wine or dry sherry. I have used all, and they are all satisfactory. The topping does not change the quality of the original wine.

There is no set rule about racking. You must use your own good judgment and consult your own wishes.

The Worship of Beauty * 67

However, here is something to be watched. Sometimes fermentation takes a long time. If you notice some tiny bubbles, let the wine stand upright, lightly corked, and allow it to complete its fermentation. If you wish to stop the process, add about an ounce of brandy or vodka.

BOTTLING AND THE PLEASURES FOLLOWING

DRINK WATER like an ox,
Wine like a King.
Old proverb

The Bottles and the First Law

The first law of bottles is that they must be just as clear and bright as the wine. One of the failures I have had during my years of wine making I attribute to bottles that were not properly cleaned. I decanted a gallon of good daisy-flower wine into white bottles. Two were perfect, the other three began to show a strange sea-green scum on the top. I threw away the three bottles. The only thing that could have caused the scum was uncleanliness of the containers. And so now, whatever bottles I use, new or old, I boil them first, dry them on a stove, and then close them with clean corks until I am ready to use them.

Bottles of any color, size, or shape will do. I prefer clear glass, because it enables me to see the condition of the wine: whether it is still fermenting and whether it needs more decanting.

Always smell the bottles before filling them. Your nose will tell you quickly whether or not they need another washing. Be sure there is no debris or dust in them.

If you wish to buy bottles, which are not expensive, you can find the names of dealers in the classified telephone directory. Some bottle dealers are listed at the end of the book.

The bottle should be filled to an inch below the opening. Remember that your wine, even when well sealed, continues its mysterious progress toward the ultimate perfection of bouquet, color, character, and flavor, and this requires just a little air.

Corks and Corking

Corks are important. Do not use old corks for bottling new wine. The cracks made by a corkscrew allow too much air to enter and may cause leakage. New corks are easily secured and are inexpensive. Be sure to find perfect, fine-grained ones, without too many air spaces in them.

I am exacting about the clarity of my wine, so even after I have siphoned off the clear wines from the containers into the bottles, I do not cork them at once. I first put in any clean tapered corks I have on hand, and let the bottles stand upright for a week or two. If there is any sediment, I resiphon and get rid of it. Then I am ready for the proper corking.

You cannot cork a bottle properly by hand; the cork is too wide and generally straight lengthwise. A simple plunger-corker (I paid

$1.25 for mine) is effective. The instrument is a long and hollow chamber just wide enough to hold a cork. One end is open, fitting the opening of a standard-size bottle. At the other end is a plunger going through the whole length.

First, put the cork into warm water for about a minute, then with your fingers put a little olive oil or any cooking oil around the end of the chamber, which has a metal ring. Set the cork into the chamber as far down as possible. Put the end of the corker over the bottle and force the plunger down with a hammer. Decrease the strength of the hammer blows when you get near the glass so as not to break it.

The oiling should be repeated after every fourth or fifth cork is set, so as to ease its descent.

If you venture into home fruit champagnes, you need heavy bottles, champagne corks, caps, and wire baskets. For, though homemade champagne is not like a purchased one, it will blow out or even smash a light bottle, just as the other will.

All the necessities for corking are procurable from the firms that sell wine-making accessories.

When the bottles of wine are corked, let them stand upright a day or two, for the corks to dry. Then lay them down and examine the corks a day or two later, to make sure there is no seepage through them. If there is, the bottle must be recorked.

When the corks are absolutely dry, I like to seal them with melted paraffin. Dissolve the wax in an old pot or tin can of any kind, then pour some of it with a spoon on the cork first, forming quite a heavy coating. When it has cooled, dip the end of the bottle one-half inch into the wax. Though this is not absolutely necessary, it is a good precaution in order to keep the wine healthy and pure.

Sometimes you may see tiny air bubbles along the neck of the bottle. That portends either dire tragedy or wonderful delight. It may mean that the wine is turning into vinegar because of some faulty step during its making. Or it may mean that the wine is *pétil-*

lant ("tiny bubbling"), which is delicious to the taste. You will find such *pétillant* wines in some parts of France. I had one in Marseilles that I have never forgotten. I also found some in a case of Yugoslav Riesling, and you may have the good fortune, as I have had, to find it in your own wines now and then.

Of course, only time and tasting will tell the tale, but from my own experience I have found that these bubbles more often portend a *pétillant* wine than vinegar.

Labeling

Labeling is a convenience and a pleasure. It is a proud moment for the vigneron, the wine maker, to bring out a bottle of wine of his own making, with the name and year enshrined in fine colors and pleasing design.

You can buy simple red-bordered labels in any stationery store and print or type the names of the wines and the year they were made. You can use simple gummed paper or ordinary paper that you glue on.

But there are some firms in England, where, as I said, home wine making is part of country and farm life, that sell labels with the names of the wines in attractive and colorful designs. These enhance the appearance of the bottle.

There is still one more type of label, which will give a most inviting, lovely-to-look-at result. You can have a special plate made with your own design and name on it, and a space for putting in the name and date of the wine. Or you can get a catalogue from a company that makes bookplates. Some of the plates do not have *Ex Libris* on them, so you can write your name, wine, and date. One catalogue I saw had some plates that were quite appropriate.

But whatever label you use, you will feel, even as I have felt, a glowing aesthetic compensation when you see a row of bottles filled with good wine nicely labeled.

If the wine is stored in a dampish cellar, put a rubber band around the label. Dampness often loosens the glue.

I really use two labels. Since bottles are laid down flat, you can tell the contents only by lifting the bottle, which is not good for the wine. Here is how to circumvent this difficulty. Cut small tabs about two inches square from light cardboard—old wedding invitations or engagement announcements are excellent. Then punch a small hole at one side and put a small rubber band through it. Write the name of the wine and the year on the tag, and fasten the loose rubber band around the neck of the bottle. Thus you can tell the contents from the loose label without disturbing the wine.

Record Book

It is an excellent idea to keep a record of the wines you have made. Any kind of alphabetic notebook will do. Put down the name of the wine, the date made, and any notes that will be of future value. Then put down a small mark (/) for every bottle you have. When you remove a bottle, put a line across the mark (X). Thus you can tell at a glance how much of the wine is left.

# THE FINAL WORK

Maturing and Storing

THERE ARE SO MANY fine thoughts and proverbs to prove the value of maturing wines that I don't know which to begin with. Shall it be with: "Wine is like a scholar; the older he gets the riper he becomes," by Yuen Mei; or should it be: "The bread of yesteryear is the meat of today," by an author whose name I do not recollect? Whatever thought or proverb, the fact remains that the wine of yesteryear is the joy of today. But alas! Time is frightfully slow for those who must wait. Nowhere does this hold so true as in home wine making, where time truly is the crowning achievement. Nowhere is temptation so great to cheat time—to taste and drink before the wine is ripe and mature. A temptation ever present, from the very commencement of the work.

I speak with authority on the subject, for I have experienced it, yielded to it, and even now—at times—am still a most willing sinner, but not too often.

Of course, you taste the changing must while it is performing its flickering tarantella during fermentation. Then you taste it when it becomes still, to correct the flavor. Here you should stop and keep in mind that only a mature wine is a good wine. But temptation is too great, the power of resistance too weak. There is also the eager, egotistic pride in letting friends taste what you made with your own lily-white hands. But all this is wrong. Wine *must* rest in peace for at least one year before you or your friends drink it. Be happy with the visual array of bottles in proud rows. The waiting will bring the magnificent reward of a mature, full-bodied, rich drink, fine for the body and soaring for the spirit.

The resting of wines cannot be overemphasized. It is truly essential for the final creation of the spirit in the drink.

When I made my first tomato wine, I tasted it and found it raw, immature, and unpalatable. One year later it had a pleasant Burgundy-dry wine flavor that was a delight. I am positive that in another year or two it will be still better. All wines, like human beings, need time to ripen and come to their full strength.

Bertha Nathan, of Maryland, a marvelous storyteller of Negro folktales, gave me a royal gift of three bottles of 1932 elderberry wine, one bottle of 1932 currant wine, one 1929 parsnip wine, and one 1926 blackberry wine. I shall not describe their rich maturity and bouquet, for that will need words I have not yet discovered. They were made down in Maryland by her parents and aunts. All homemade wines can achieve the same success, if only you will give them a chance. You must learn to do this, in justice to the wine and to your palate.

With the fine wines, Bertha gave me some fine anecdotes of the old folks who made them, and some of them are well worth repeating:

"Solitary drinking was called 'wardrobe drinking.' . . .

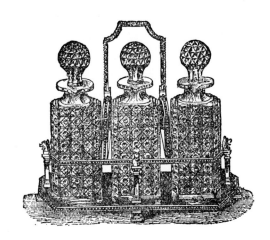

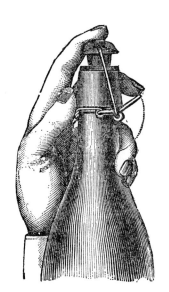
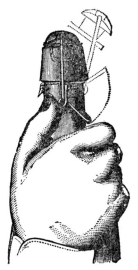

"My grandmother once accidently served well-spiked lemonade to a teetotaler. Said the lady, 'Clara, this is delicious lemonade; and please don't tell me how you made it.' . . .

"There was a saying—grapes were the best for wine: you could make an arbor to sit under, thereby helping the outer man; and you could make wine, thereby helping the inner man. . . .

"The old way of making brandied peaches was to put ten peaches at a time in hot lye. Then take them out, let stand to cool, and rub off the fuzz with a coarse towel or burlap. One time William took a peach out, too hot to handle, and used bad words. Said the mistress of the house, 'Do you think the preacher would use those words in church?' Will thought for a few moments, then he said, 'Yes'm, only he'd have to 'range 'em a little differently.' . . .

"The Captain gave the old man a bottle of just so-so wine and then asked him how he liked it. Said the old man, 'It was jes' perfect.' 'I am surprised,' said the Captain; 'I thought you knew drinks better than that.' 'Yes, suh, I does. But I still say it was perfect, because if it had been any better'n it was, you wouldn't have given it to me; and if it had been any worse'n, I couldn't have drunk it.' . . .

"The preacher said he had never tasted wine. So the women decided it was time he did. When it came time to send their September offering of fruit, they placed in the center an unmarked bottle of grape wine. When the preacher thanked the women for it, he said, 'It was just beautiful to look at; grand to taste; but the memory of the spirit in which it was given will always linger on.' "

So ran some of the tales of Bertha Nathan. But now let us return to our wines.

If you have a home in the country, find a cool place in your cellar or a room that is not heated, and lay the bottle down so that the liquid is in contact with the cork. It is best to make a closet with shelves at a slight tilt, 5 ½ to 6 inches apart and 16 inches deep.

I have two such closets in my cellar, with room for two hundred and fifty bottles. Of course, you don't have to make them as large. When you put the wine on the shelf, put down on a sheet of paper the name of the wine and the year it was made and the number of bottles. Keep that under the necks of the bottles. Number the shelves: 1, 2, etc.

Next, put down in your record book the number of the shelf and the same information about the contents of the bottles. This gives you a double record.

If you live in a city and have no special room for your wines, then you might buy metal or wooden wine holders in a department store, but these are quite expensive.

The best place to store your wines in the city is in a bedroom or study where there is no radiator, or where the radiator is shut off all the time. I have such a room in the city, and have in it over two hundred bottles of French, Yugoslav, and Spanish wines, and my own flower and fruit wines.

Serving Wines, and When to Drink Them

A word about serving those wines so beautifully laid out in your bin, either in your country or city home.

If you are adventurous in gustatory pleasure, you will use your fruit and flower wines exactly as you do your fine grape wines, and you will go through the same exciting taste thrills.

There are sweet wines: red, crimson, rose-colored and yellow; and there are dry, tangy wines in varying colors from blood-red to straw-white. Wines are suitable to accompany festive fine food and also for afternoon "tea" or after-dinner conversation. There are homemade wines, including aperitifs such as rice wine, suitable for all occasions. I like to serve rice wine hot, in small jade cups of many green and brown hues.

The bouquet and flavor of these wines brings you into a new world of taste pleasure. Thus, a dry parsley or tomato wine or an old dryish raspberry wine will add a new, exquisite taste to a well-prepared chicken or a sautéed-and-sesame-seed-covered fish. An old courtly elderberry wine that has lost its sweetness and has its own rare taste well accompanies a roast or a crackling duck. And a cooling green, sweet mint wine or a pink rose-petal wine makes of your cake a royal Lucullan feast. You develop a new sunlit world of rare, gay, gustatory and culinary luxury.

When should wines be served? On every occasion when you want a little pleasure. Here, too, classic patterns should be observed: amber-colored wines, commonly called "white," should be chilled; red, served at room temperature.

Many of the fruit and flower wines will be on the sweet side—though remember, they can be dry as well, if you don't add too much sugar. I call sweet wines "occasional" wines; they are especially suitable for a leisure hour in the afternoon, with dessert, or to sip while reading. Fruit and flower wines, to many of which folks attribute medicinal powers, come in for much usage. I myself have found that after three hours of steady writing, when my head is a little heavy or when I have an actual headache, a glass of dandelion wine brings relief as well as pleasure. I also take a glass of it before bedtime and find that it helps to induce good sleep and pleasant dreams.

Usually I follow the classic dictates: white dry wine—a rhubarb, peach, pear, elder flower—with light meats or fish; with red meats, a red wine of elderberries, blackberries, blueberries, or raspberries.

But I also believe that every person should drink his favorite wine, irrespective of classical rules. For whichever one you drink will enhance the pleasure of the food and bring a feeling of warm tingling to the mind and body. It will stimulate thinking and conversation.

The only time wines may be mixed with water is on a drowsy summer afternoon when you want a cooling drink with pleasant vinous echoes. Then you may mix a white wine, or any fruit or flower wine, with charged water. The homemade wines are particularly appropriate for such a drink, because of the addition of the fruity or flowery flavor and perfume.

And finally, remember that your good homemade wine will do for you what Proverbs in the Great Book says it does for man: "Give . . . wine unto those that be of heavy hearts. Let him drink, and forget his poverty, and remember his misery no more."

This is a truth from time immemorial.

Of Glasses

No sensation or experience has but a single cause; it is, rather, the result of a multiple cascading series of facts and feelings. Drinking a homemade wine is not just swallowing a nice-tasting, strong liquid; it is the expression of endless facets of life, past and present. The table is set with a bright tablecloth; there are bright shining plates, gleaming silverware, and crystalline glasses. It is a gala occasion and as such needs its own special dressing.

Just as fine clothes are an additional pleasure to a festive hour, wine, in special clothes—the glasses—adds sparkle to the festive luster. And somehow, particular glasses fit each wine. I can't conceive of a deep red rose-petal wine in a dark green glass. Gaudy glass never fits good wines, anyway. A good wine has character and strength and should never be poured into a fancy rococo glass.

For your own good homemade wines, as well as all good wines, use only pure, clear, white glasses—crystal, if you can get them and can afford them. They are lighter, more pleasant to the touch, and beautifully bell-toned when sounded.

It is good to have them with long, thin stems that will balance well in your fingers.

They should be somewhat bell-shaped, narrow at the mouth. Never fill them more than two thirds full; even a little less is better, thus giving the aroma, the perfume, the bouquet, a chance to fill the empty space to complete the trinity of the joy of wine drinking:

with the eyes, the nose, and the mouth. This is the only way to enjoy a wine of rose petals, or spiced geranium leaves, or peaches, or elderberries, or rice, or dandelions, and all the rest of the galaxy of wines made from fruits of the trees and the earth.

Since I mentioned rice wine and I am on the subject of drinking vessels, one final word, for thereby hangs a world-wide tale.

I am particularly fond of warm rice wine—saki. It goes perfectly with the reading of Chinese and Japanese poetry, and as an aperitif. So I decided that I wanted to drink the warm amber-colored liquid from multishaded green-and-gray Chinese jade cups and poured from an old carved Chinese jade bottle. For six years I searched throughout the world for them. When I did see some, the price was far more than I could possibly afford. But I believe that luck is always by my side—provided I encourage it with persistency. One day I found exactly what I had sought so long: twelve jade cups of exquisite beauty in spinach-greens and gray-browns, delicate as Chinese songs. And there was a wine bottle of the same indefinable lovely colors, with a carved inscription of good luck and happiness.

I am willing to take an oath—and I know my friends who are well versed in the beauty of wine and who share my evenings of perfect dining pleasure will agree with me—that the delicate-scented warm rice wine tastes more delicious and tangy for the beautiful jade bottle from which it is poured and the exquisite cups from which it is drunk.

FRUIT WINES

 *Apple Wine*]

"AS THE APPLE TREE among the trees of the wood,
  So is my beloved among the sons.
I sat down under his shadow with great delight,
  And his fruit was sweet to my taste."

The words are from one of the most beautiful songs ever written. But the apple tree is not only deep-rooted in the Bible, it is deep-rooted in all the lore of the world. And probably there is no fruit around which more controversy has raged, ever since the Biblical story of creation. The thorny question was, and still is: Was it, or was it not, the tree and the fruit of knowledge in the Garden of Eden, the eating of which brought drenching misfortunes to the world?

There are oceans of nays and as many ayes. Endless other fruits and trees have been set in that Elysian retreat where there was joy and peace. The fig tree, the pomegranate, the orange, the banana, the apricot, the palm, and others claim the same honor, or have had thrust upon them the blame and shame.

The nays have it that the early Renaissance Olympians of art were the first to select the apple for this dubious honor. But it was not until Milton's marching paradisiacal poems that it became a truly popular tradition.

Perturbing controversies have followed the apple through the world. From it have sprung great tragedies; it has had more prominence and power in mythology than any other fruit. And the same holds true in fairy and folktales and in literature in general.

Probably it is the most mystical of fruits; it has the attributes of marvelous powers. The way in which this fruit twines its way around the whole world of man speaks words attuned to deeper meanings of life.

The Greeks, in particular, had the fragrant apple as the core of many of their violent loves and bloody tragedies. The possession of an apple tree in the days of Dionysus gave the owners the godly gift of supernatural powers. When Hera and Zeus were wedded, the lovely goddess' gift to him was the famous golden apple. Every schoolboy knows the celebrated race that Atalanta lost for three golden apples. And all the world has rung for centuries with the immortal tale of Troy, precipitated by a golden apple. It was Paris who gave to the goddess Aphrodite the coveted apple, ultimately bringing about the downfall of a city and creating a tale that will be told as long as men tell tales.

There was Tantalus, ever striving to reach the apple to lighten his tragedy. There is a host of tales in like vein.

In the cold North, bards and priests wove around their hard apples a tale sharp and clear as their white, iced air. The eating of the perfumed fruit kept the gods eternally young.

Perhaps a Scandinavian Voltaire originated the proverb: An apple a day keeps old age away. The Scandinavian goddess Iduna was identified with the tree of immortality, which was the apple tree. She kept a box of apples that the gods ate when they felt age approaching, thus renewing their youth.

In many lands it is an accepted belief that the apple tree makes women fruitful. Women of many peoples, like those of Kara Kinghiz, roll themselves under a solitary apple tree so as to have as many children as the tree has apples.

It is noteworthy that, according to the medieval alchemists, the gleaming phoenix was the only bird that refused to eat the apple, the mystic forbidden fruit. Eve, with the desire common among evil-doers to share black deeds, offered to all the fruit she was forbidden to eat, and all ate except the golden bird, and that is why it has remained immortal.

There is a charming saint's tale, with the apple equal to the heroine. It happened in the days of the Romans. Dorothea, a lovely maiden who had been converted to Christianity, refused resolutely to worship the Roman gods and was therefore condemned to death. Theophilus, a renowned Roman lawyer, must have loved her, for she was exquisite to behold and rich in rare sensibility. But she was steadfast in her belief and ready to give her young life for it.

A milling throng lined the roadway when she was led to the execution. Theophilus was there, too. The leering crowd jeered and derided her and shouted spewing words about her new God.

Theophilus, carried along by the mob, cried, "Send me some fruit from your new heaven, Dorothea!"

"As you desire, Theophilus," she said calmly and sweetly.

She asked the sworded guard's permission to halt, and fell on her knees and prayed to the heavenly Father. As she arose, there suddenly appeared beside her a boy with golden hair, holding a basket of lovely flowers in his hands. On top of the flowers lay three apples of exquisite beauty with a rare, redolent perfume.

"Give the basket with the apples to Theophilus," she said to the fair-haired boy, "and tell him they come from the Paradise of our Lord, where I hope to meet him some day." Then she went off between the armored soldiers.

Theophilus remained behind, holding the basket in his hands, from which came the heavenly apple perfume. The sight and scent entered his body and his heart and his mind. He ate the fruit and, in time, the miracle of the apples showed him the truth of the new faith and he embraced the new religion, even as Dorothea had done.

In the early centuries, ripe apples were used in cosmetics. During Queen Elizabeth's days, the most popular face cream was prepared with mashed apples, to which were added rose water and hog's grease.

A celebrated complexion paste was made of mashed apples, crushed almonds, bread crumbs, equal quantities of rose water, white wine, and a little soap. "This you must cook until it is a fine, smooth paste." That surely must have softened the Elizabethan ladies' skins to make them a delight to the lips, and whitened them to a feast for the eyes.

Who of us has not read the Russian fairy tale of the great, gleaming, golden bird that came in the radiant moonlight to steal the golden apple in the Czar's garden?

And there is the tale of William Tell, the mountaineer of Switzerland, shooting an apple off his son's head as a cry of defiance to tyranny.

Divination, too, was an integral part of the apple. Squeeze an apple pip between two fingers and note in which direction it flies—there waits the lover.

Folks like you and me in England used to designate the sunny days when perfumed apple blossoms were in their richest pink-and-white glory as "sweet Sunday."

The mistletoe, a sacred plant, grew often on the apple tree, and so it was worshiped by the Druids. To the English the tree also sym-

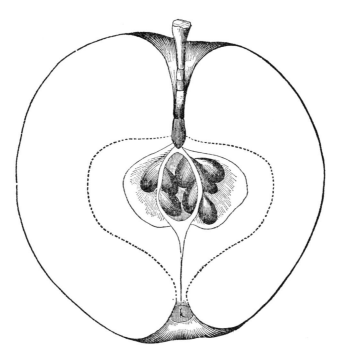

bolized manhood, and often young folks danced around it instead of around a Maypole in the fresh month of May, sprinkling the tree with apple cider, a custom dating back to dim and ecstatic mysteries.

In Persia, the apple is the fruit of immortality—an echo of the Bible, and a belief somewhat similar to the one held in Scandinavia.

In the United States, too, the apple has played an important part in lore and life. Emerson, the good philosopher, called it the "social fruit of New England," for around it centered numerous social activities. "Apple cuts," "apple parings," "apple bees," and apple cider added to the colorful social life of colonial days. And here I cannot resist mentioning our adopted national dish of apple pie. Those made in my own home should be inscribed in a special gilded chapter, for they deserve the Biblical words: "Comfort me with apples." If you also want a taste of Paradise, here is how to gain it: by means of a pie made according to the recipe of my wife, a great culinary artist.

First, be sure to put cinnamon into the dough. Then bake the bottom crust slightly to eliminate the morassy sogginess so often found in the bottom crust of pies. Cut the apples into nice chunks, put them on that bottom crust, sprinkle cinnamon generously all around, add a touch of lemon juice to sharpen the flavor. Next, take a few spoonfuls of good elderberry wine, or Malaga, or rich Madeira and sprinkle it over all. Then top with a few marshmallows. Put on the top dough, which must also contain a little cinnamon, and bake it brown. Serve it HOT. Then write me about it. If you don't think as I do, I will eat the metal pie plate!

It is interesting to learn that there are records of pies from as far back as the fourteenth century. Recipes for them have come down from those days.

Your grandmother, I am sure, spoke of "comfort apples," apples studded with scented cloves. Such apples were often brought as gifts, as we bring flowers or candy today. And, of course, there was apple butter. It was, and is, delicious. It is a great pity that it is not popular today. Here is an old recipe for making it, and if you will take time off from playing bridge or watching TV, I promise you a greater pleasure than either of these two can offer.

Get a gallon of pure cider and boil it down to about two thirds its volume. Then add about five pounds of sliced apples, one-third gallon dark molasses—which is an old, fine folk remedy for many ailments—and finally, spices you favor: cinnamon, mace, ginger, and just a touch of pepper. Use your good judgment with these to suit your own taste. Let it boil very gently for twelve hours, stirring it as often as you smell it. Finally, put it into sterile jars. In Pennsylvania it is called apple butter, in New England I have heard it called apple sass.

The parings of apples used to be thrown over the shoulder to see what kind of letter they would form. That was the initial of the coming lover.

The only saint-hero in American lore is blood brother to the apple! Johnny Appleseed spent his ripe years planting apple seeds over the land with loving and unselfish care.

There are numerous tales about the fruit in almost every state. Here is one I just heard in Oswego, New York. There old Alveh Walker, for black reasons and in a black moment, cursed the apple tree in his back yard and it never bore apples again. So, you see, the apple tree can be willful, too.

There are also innumerable games in which apples play an important part. In the olden days, the trick was to bite an apple in the wassail bowl. Bobbing for apples on eerie Halloween is a common game even today. And there are still some young girls in the remote, winding corners of our land who take an apple in one hand, a lighted candle in the other, and eat the fruit in front of a mirror, hoping to see the future love in it. That custom was quite common in Scotland, whence it probably migrated to the New World.

The number of culinary recipes in which apples are used is

legion, and of course there are many proverbs centered around the apple. Interestingly enough, they often revolve around the mystic and medicinal powers of the fruit.

"Eat an apple going to bed,
Make the doctor beg his bread."

or

"Eat an apple going to bed,
Knock the doctor on the head."

Thus, there is no recess of man's life anywhere in the world where the apple has not played a rich part of folk life. Add to this the fresh, invigorating taste of the flesh of the fruit, the clean savor in mouth and body after eating it, and it is small wonder that apple wine and apple drinks have been popular from time immemorial.

I am not concerned here with the making of cider, which is not a wine. But I will tell about kwas (the "a" is pronounced as in the word "far" and the "s" as in "sip"), which contains a small amount of alcohol and which I used to buy in my native city in Austria. It was a cool, soury drink with small, brownish, liquid-soaked apples floating on the top, sold in the busy market place.

And where do you think I found it in our own land? Just about a mile away from my farm in Putnam County, New York. There it is made in a White Russian hermitage by long-cassocked, long-bearded Russian Orthodox priests, aided and abetted by Robert Le Gros, who has a most amazing knowledge of anything and everything.

The Recipes

Somehow it is difficult for me to accept an apple wine. Perhaps the early life association with the apple drink of kwas and the later years with cider have set the apple apart from wine. Yet apples make an excellent wine with a faint, cidery, champagne tang bringing a very refreshing flavor. Let me state here that my first trial at making apple wine resulted in complete failure, but the subsequent attempts were crowned with royal success. It was a good, strong wine with the champagne-cider zip of which I have already spoken.

It is a good wine to add to any of those whose flavors are not quite pleasing. I added it to a barley wine that was somewhat harsh. The addition of the apple wine toned it down and made it a pleasant drink.

Here are five different recipes. You can use any kind of apples you like.

I

The simplest method is to make the wine from fresh apple juice that contains no preservative. This can be bought from a country cider press, or from roadside stands where they make their own apple juice, or from apple orchards where they often make their own juice. You can, of course, buy apples and press your own fresh apple juice.

You need: 2½ gals. apple juice
2 lbs. sugar
½ oz. yeast (2 packages)

1. Put the apple juice into an enamel pot.
2. Add the sugar and stir it with a wooden spoon until it dissolves.
3. Bring to a boil and let it boil gently, removing any scum that rises. When there is no more scum (after about half an hour's boiling), pour into a stone crock and let it cool.
4. Dissolve the yeast in ½ cup warm water and pour it into the crock.
5. Cover; let it ferment from fourteen to twenty-one days, or until the visible fermentation has stopped. If any scum rises to the top, remove it.
6. Strain; put it into half-gallon or gallon glass jars with wide

mouths. Let stand for at least two weeks. If it is not clear by then, fine with eggshells (two eggshells for each gallon) or isinglass (1 spoonful of dissolved isinglass to each gallon). When clear, bottle and cork.

II

You need: 16 lbs. sound apples
2 gals. water
½ oz. yeast (2 packages) (*this may not be needed*)
2 lbs. sugar for each gallon of liquid

1. Wash the apples thoroughly. Cut them into small pieces, but do not peel them or remove the cores. Put them into a 4-gallon crock or into a perfect, unchipped 4-gallon enamel pot or pail.
2. Boil the water. Cool; pour it over the apples.
3. Cover the container and let it stand in a warmish place (65° to 70°).
4. If, after a day or two, fermentation has not started, dissolve ½ ounce yeast in ½ cup warm water and pour onto the must. This is rarely necessary; the apples will usually ferment without yeast. Let stand for ten days, until the strong, visible fermentation has ended.
5. Strain through a hand towel or muslin into an enamel pot. Clean the crock of any residue, and pour back the strained liquid into it. For each gallon of liquid add 2 pounds sugar. Stir it well with a wooden spoon until it is completely dissolved. Cover well again and keep in a warmish place for a week to ten days, taking off any scum that rises to the top.
6. Strain and clear as in the other recipe.

If you note that the wine is still "playing" (fermenting), which you can tell by tiny bubbles on the sides of the bottles, put in the corks very lightly until the fermentation stops. A teaspoonful of very dry sherry, brandy, or vodka will hasten this.

III

You need: *16 lbs. apples*
2 gals. water
2 lemons
3 oranges
½ oz. yeast (2 packages)

1. Wash the apples, cut away any damaged parts, and cut into pieces. Do not remove the skins or cores. Put into an unchipped enamel container.
2. Pour the water over the apples and bring to a boil, allowing the water to simmer until the apples are quite soft but not mushy.
3. Let the juice rest for a day, and then siphon off to eliminate the sediment.
4. Wash the lemons and oranges. Cut off the rinds thinly and put them into the liquid. Squeeze the fruit and add the juice to the other liquid.
5. Warm ½ gallon of the liquid and pour back into the crock so that all the liquid is lukewarm.
6. Dissolve the yeast in ½ cup lukewarm water and pour it on the must.
7. Cover well and let ferment from seven to fourteen days.

    Then follow the procedure described in the previous recipes.

    Here is a pleasant suggestion for your apple wine. Serve it very cold, like champagne, and when you open the bottle put in a lump of sugar and shake until it is dissolved. It effervesces, becoming somewhat *pétillant* and champagny.

IV

You need: *16 lbs. apples*
2 gals. water
½ oz. yeast (2 packages)
1 lb. sugar

1. Wash the apples, cutting away the bruised parts. Cut them into small pieces and mash them by hand or with a wooden mallet or in a press. Put the mash into a crock.
2. Boil the water and pour it, boiling, over the apple mash.
3. Cover well and let it stand for about two weeks in a warmish place (65°-70°). If fermentation does not start in two days (most of the time it does), add ½ ounce yeast. Fermentation will take from seven to fourteen days.
4. Strain the liquid and add the sugar. This will result in a nice, tartish wine. Be sure the sugar is completely dissolved.
5. Clear, following the previous directions. Then bottle.

V

You can also make apple wine from good, fresh, unadulterated cider. Here is the recipe.

You need: *any amount of cider, without preservatives*
sugar, 2 lbs. for each gallon

1. Put the cider into a crock.
2. Add the sugar, dissolving it thoroughly.
3. Leave open for two to four days.
4. Pour into gallon jugs, corking them lightly.
5. Let them remain still for at least a year, then your wine will be ready.

Practically every kind of apple wine can be made *pétillant* by the addition of sugar, as described in the third recipe.

Apricot Wine]

EVEN AS THE ROSEATE DAWN and the bright sun go together, so do apples and apricots—Biblically. It is mainly around these two that the great controversy has raged: Which was the tree of knowledge in the Garden of Eden? Many insist that it was really the lovely apricot that brought trouble to Adam and Eve and to the world.

The great Biblical scholar, Tristram, claims: "The apricot is the most abundant fruit in the Holy Land and meets all the requirements of the context, and is the only tree that does so." There ". . . it flourishes and yields a crop of prodigious abundance."

In Persia, the apricot is called "the seed of the sun" because of its loveliness of color and delicacy of scent. And that you will surely believe if you look at and eat apricots ripened on dark trees by warm sunshine.

The Chinese not only hold the apricot in deep veneration, but consider it the tree of prophetic and oracular power. The great sage, Confucius, completed his commentaries on the ancient Chinese books under an apricot tree. Another celestial honor brought to the apricot tree was by Lao-tse, the great Chinese religious philosopher.

One day the mother of Lao-tse, before he was born, wandered outside her village and finally sat down under a pink blossoming apricot tree, and there took place the miraculous birth of her son from the side of her body. He was born white-haired and filled with wisdom. He spoke from the moment he was born. He turned to his mother and said, "I take my name from that tree," and thus he was known and renowned.

The folks of that land made a wine called "apricot gold," the drinking of which was said to prolong life to seven hundred years.

An apricot wine, if well made and well matured, will also find honor among those who appreciate fine homemade wines.

The Recipes

Here are two recipes for making apricot wine. From this fruit, as from peaches, you will have a dryish wine, approximating a sauterne.

In these recipes you can use either ripe or dried apricots: one quarter to one fifth the weight of dried fruit as compared to the ripe. Try to buy the dried apricots in stores that sell health foods. There you will generally find dried fruit to which no sulphur has been added and for which natural fertilizers have been used, and hardly any sprays. In any case, it is wise to wash any fruit thoroughly before using it. Dried apricots should be allowed to soak in water until they swell up.

I

You need: *2 gals. water*
3 to 4 lbs. sugar
15 to 25 lbs. fresh apricots
½ oz. yeast (2 packages)

1. Put the water into an enamel vessel holding 4 gallons.
2. Dissolve the sugar in the water.
3. Take ripe, sound apricots (the more apricots used, the richer the wine), wash them thoroughly, cut them into small pieces, and add them to the water.
4. Break up about half the apricot pits and put pits and the kernels into the water.
5. Add another quart of water and set the whole mixture on the fire. Keep it boiling from one to two hours.

APRICOT

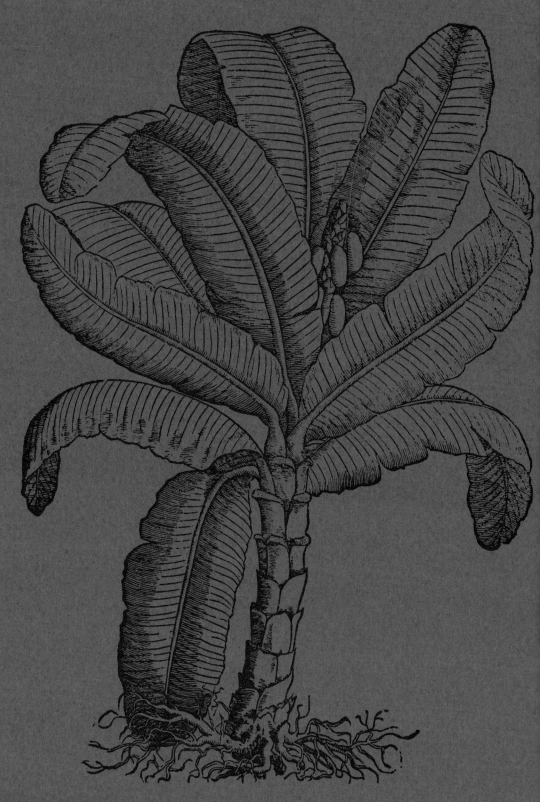

6. Strain the liquid through a cloth into a crock and let it cool down.
7. When it is lukewarm, dissolve the yeast in ½ cup warm water and pour it in.
8. Cover the crock, put it in a warm place, and let it ferment. This will take from ten to twenty-one days. When the violent fermentation has stopped, strain the wine off into glass jars or gallons and let it rest for a week or more to clear. If it is not clear by then, fine it, rack it again if necessary, and then bottle it.

II

You need: 2½ gals. water
3 lbs. rice or barley
2 lemons
3 lbs. dried apricots
4 to 6 lbs. sugar
½ oz. yeast (2 packages)

1. Put the water in an enamel vessel.
2. Rinse the rice or barley thoroughly and add it to the water.
3. Add the thinly cut rinds of the lemons.
4. Boil the mixture for a full half hour, then strain it into a crock.
5. Remove the cereal (you can eat it if you wish), clean the vessel, and pour the liquid back into it.
6. Wash the apricots well, put them into the liquid, and let it rest until the apricots have swelled up. When they are soft and full, bring the liquid again to a boil and let it simmer for a good half to three quarters of an hour.
7. Strain into the crock and add the sugar and the juice of the lemons. Dissolve the sugar thoroughly.
8. Let the liquid cool; when lukewarm, add the dissolved yeast in the usual manner, either with or without toast.
9. Cover, set in a warm place, and let it ferment, which will take two to three weeks. Follow with clearing, as in the preceding recipe.

 Banana Wine]

HERE AGAIN IS KEEN competition for the honor of the controversial tree of knowledge that the Lord planted in the center of the Garden of Eden. The Jews and the Greek Christians who inhabited Syria believed, and the people of Honduras still believe, that the banana was the tree from which Adam, at Eve's behest, ate the fruit. The banana was the most favored and most beautiful of all trees in the Garden.

"But the Lord told Adam and Eve that they must not eat the fruit of it, or even touch it with their hands, for if they did, they would die. Then one day there came that mean, old, double-faced snake . . ." And the world knows the rest of the sad, sad story.

And so the first clothes of man were not fig leaves or apple leaves but banana leaves. So say the Honduran folks. And the leaves are big enough to be used as umbrellas in the torrential rains of Indonesia and wherever they grow.

John Lindley, the famous English botanist who wrote many volumes on plants, also remarks that the banana was thought to be the tree of knowledge; hence one of its names is *paradisiaca*.

India considers the banana tree and the fig tree the Trees of Life.

In central Africa there is a charming, innocent belief that any nice female on whom a purple banana flower falls while she is walking or working will become pregnant and bear a child. So any lady accused of infidelity by a hard-hunting husband would be honorably acquitted by pleading that a banana flower had touched her body.

Whatever folks say of the banana anywhere in the world, none can deny that it is a very tasty, succulent, meaty fruit; that it is a healthy and useful fruit; and that it makes good wine. Here is the recipe.

You need: *2 gals. water*
8 to 12 bananas (depending on their size)
juice of 2 oranges
1 lemon, sliced
4 to 6 lbs. sugar
1 slice toast
¾ oz. yeast (3 packages)

1. Put the water into an enamel pot.
2. Slice and add the bananas. There should be 2 pounds of peeled bananas. I have found that five ripe, medium-sized bananas make 1 pound. I therefore use ten bananas to 2 gallons of water. More will give a richer and heavier wine.
3. Add the orange juice to the water.
4. Add the washed, sliced lemon.
5. Bring the water to a boil and let it simmer for about twenty to thirty minutes.
6. Strain it carefully into a crock. Do not squeeze the fruit.
7. Put in the sugar and dissolve it thoroughly.
8. When the liquid is lukewarm, prepare a piece of toast and dissolve the yeast in ½ cup warm water. Spread the yeast on both sides of the toast and pour the rest on the liquid in the crock. Or you may use the yeast without toast, as in other recipes. Cover the crock and set it in a warm place (65°-70°) for fermentation. It will take from ten to twenty-one days.
9. When the visible fermentation has ended, strain the wine into glass containers. You will find it quite cloudy. Let it stand at least two weeks. If it does not clear, fine it with isinglass or eggshells.

If it continues cloudy, clear it through filter paper. I have never known that to fail.

You can create a slightly different flavor by adding 1 pound dates, pitted and cut up, at step 4.

[Cherry Wine]

STRANGELY ENOUGH, there is nothing I could find about cherries in the land of lore, apart from Washington's famous youthful tale, yet I have a feeling that there must be some choice bits I have overlooked. I vaguely remember reading somewhere of the Countess of Desmond in the days of King James I, at the age of 140 (she may have been 145), who climbed a cherry tree and fell off it to her untimely death. But of course cherries, even as roses, are a frequent literary simile to describe lips and cheeks.

And so, for the nonce, I can turn to—

Recipes for the Wine

I

You need: 10 lbs. cherries, any kind: sour or sweet, yellow or red
2 gals. water
1½ to 3 lbs. sugar per gallon of juice (see step 4)
½ oz. yeast (2 packages)

1. Wash the cherries thoroughly and then leave them to dry or rub

them dry with a towel. Put them in a crock. Mash them well with a wooden spoon or with your hands. If you like a nutty flavor, break up half a fistful of the pits and add them to the fruit.
2. Cover the fruit with the cold water and let stand for two days, stirring it a few times each day.
3. On the third day, strain off the juice, pressing it through a heavy cloth. Throw away the cherry pulp, but return the pits to the liquid.
4. Now add the sugar. The amount will depend on how sweet the cherries are, and how sweet a wine is wanted. I generally add 1½ pounds sugar to each gallon of sweet cherry juice, and 3 pounds sugar to a gallon of sour cherry juice. I like a dry wine, but if you like it sweet, add another half pound or pound of sugar to the amount I suggest. Whatever amount you put in, dissolve it well.
5. Heat the liquid to lukewarm.
6. Dissolve the yeast in ½ cup warm water and add it to the liquid. Cover and set in a warm place (70°). Let it ferment; this will take from ten to twenty-one days.
7. When fermentation has ended, strain, fine if necessary, decant, and bottle.

The wine should not be drunk, tempting as it will be, for at least ten months; but it is best to drink it all within two years, since the cherry flavor is evanescent.

II

You need: *10 lbs. cherries*
2 gals. water
2 to 4 lbs. sugar for each gallon of liquid
2 lbs. raisins
1 slice toast (optional)
½ oz. yeast (2 packages)

1. Wash the cherries well and put them in the water in an enamel pot.
2. Bring to a boil and let boil for fifteen to twenty minutes.
3. Pour the liquid and the cherries into a crock.
4. Now add the sugar, the amount depending on how sweet you want the wine to be.
5. Add the raisins, cut as small as possible.
6. If you like a nutty, slightly bitter flavor, you can take half a fistful of the pits, crack them open, and put them back.
7. Dissolve the yeast in ½ cup lukewarm water. If you use toast, pour the yeast over it. Add both to the liquid. Cover, set in a warm place, and let it ferment for ten to twenty-one days. During that time, stir the must once each day with a wooden spoon.
8. When the visible fermentation has ended, strain into glass jars or gallons, squeezing every drop of juice from the pulp. Fine, if necessary, and decant until the wine is a brilliant ruby-red color, then bottle and cork.

You can drink this wine within eight to ten months, and a delicious wine it will be, dry or sweet.

III

I am fortunate in still having many close friends who believe in the real value of folklore. Bertha Nathan, of Maryland, is among them. She has helped me a good deal in this work with recipes and fine folk wines made years ago by her own family.

Apparently good, ripe cherries are as difficult to buy (and as expensive), down in Maryland as up in York State, therefore they use dried cherries instead, which are less expensive.

The first recipe, which is Mrs. Nathan's, also uses dried prunes, which I have found excellent, too.

I am giving the recipe to you exactly as she gave it to me.

You need: 1 lb. seedless raisins
¾ lbs. dried cherries (from California)
12 prunes (pitted)
2 lbs. brown sugar
1 gal. water
¼ oz. yeast (1 package)

Grind all the fruits together. Put them in a crock with the sugar and water. Dissolve the yeast in a little warm water and add it. Stir once daily for two weeks. Strain through a wire strainer and several thicknesses of thin cheesecloth. Bottle.

I inquired about the "grind all" and learned that it was done with a mortar and pestle. The yeast is dissolved in warm water and put into the lukewarm liquid. This is a very simple recipe and is easy to follow.

IV

This recipe was given to me by a woman in a small restaurant in the Adirondack Mountains where I was having lunch.

You need: 2 lbs. dried wine cherries
6 lbs. sugar
2 gals. warm water
½ oz. yeast (2 packages)

Put everything into a crock and let it stand three to four weeks, stirring once a day. Strain and bottle.

Simple enough, and it makes a good wine.

V

When I traveled through France, collecting folktales, I heard from country folks that they had a fine wine with a cherry flavor. I was interested and began making inquiries, and soon I learned how it was made.

They take enough sour, pitted cherries to nearly fill a 5- to

10-gallon keg. They crack about one quarter of the pits and put those in.

The keg is set at the back of the kitchen and the liquid is allowed to ferment and rest for two to three months. After which time, the wine is drunk.

It is a lovely rosy color and is drunk like any *rosé* wine. The one I tasted was refreshingly light and fruity. I was told at the farm on which it was made that it was excellent for people with kidney trouble.

 *Date Wine*]

HERE IS A FRUIT fabulous and rich in custom, tale, history, and life—including taste.

As a wine, there are records of it in the valley of the Indus, three thousand or more years ago. With this wine, a few thousand years more or less does not seem to matter. Some Biblical students claim that the date palm was *the* forbidden fruit in the Garden of Eden. How many fruits have contended for that unhappy distinction!

It is related that when God molded the form of Adam from the earth, some of it clung to His divine fingers. He rolled it between His hands and thus created the trunk of the date palm.

And there is the sumptuous ancient Hebraic legend of the creation of the palm tree. Soon after the world was made, a bird, big as heaven and golden as light, came up from the green earth, flying to the burning sun to rest. Up and up sped the giant bird, speedier than lightning, and when he came near the sun one of his magnificent feathers fell off his great wings. It lazed its way slowly through the air and after a time fell deep into the rich, black earth, taking on a new life and forming the roots from which grew the date palm tree.

To the Greeks, this tree with the sweet fruit stood as a symbol of victory, light, fertility, and worldly riches. It was dedicated to Apollo, the god of youth, beauty, and poetry. Poets declaimed that it was like the sun in heaven.

The Israelites sang that the palm tree was a symbol of grace and beauty, hence women were named for it (Tamar). There is a charming belief in the holy books of the Jews that the female palm tree (the trees have separate sexes, like man) wept for her male lover until a branch of him was brought to her.

The Babylonians, also aware that palm trees were of different sexes, hung the male flowers on the female trees.

There are innumerable Eastern proverbs dealing with the date palm.

"Bless the date palm trees, for they are your aunts."

"Better a handful of dates and contentment therewith than to own the Gate of Peacocks and be kicked in the face by a camel."

And in other lands: Cleopatra, the incomparable, ate the fruit; and it was buried with the great Egyptian rulers.

According to Strabo and Pliny, the ancient Babylonians drank an intoxicating date wine, and so did the Israelites during Biblical days. Herodotus had a deal to say about dates and, of course, about date wine. He wrote of the palm trees, which gave bread, wine, and honey, bearing out the statement of the Arabs that a palm tree has as many uses as there are days in the year.

The Arabs held that Allah had created this heavenly gift for the Believers and the Faithful, thus giving birth to their consuming lust to conquer all lands where date palms grew. The tree became a theme of poetry and lore and legend.

The Christians, from the earliest days, found many meanings

in the tree and the fruit. Palm branches are symbolic of Jesus's triumphal entry into Jerusalem before the feast of the Passover.

In Mexican mythology, the Tomagua Indians say the date was the founder of the human race after the Flood.

Palm leaves and branches hold an important place in religious history, and waving of palm leaves on Palm Sunday is a sign of worship and adoration.

In the fantastic medieval medical period, folks used the date to heal sunstroke, to avert lightning, to drive away mice from the granaries and fleas from the body. The flea-bitten victim had to put a palm leaf behind the Virgin's picture on Easter morning and cry out: "Depart, boneless animals," and the fleas would flee far away.

The Recipes

I

You need: *8 lbs. pitted dates (inexpensive ones)*
2 ¼ gals. water
1 handful date pits
4 lemons
3 oranges
1 to 2 lbs. sugar (optional)
¾ oz. yeast (3 packages)

1. Clean the dates of any black spots. Cut up the fruit with scissors or a knife and put it into an enamel pot containing the water.
2. Add a handful of date pits. This will supply tannin, which helps the yeast to work.
3. Wash the lemons and oranges. Peel the rinds very thinly and put them into the water. Then squeeze the juice from the fruit and add it.
4. Heat very slowly to a boil, then let simmer for a good four hours.
5. Strain the liquid off into a crock and let it cool.

6. Dates have a great deal of sugar and need no added sugar. But if you want a sweet dessert wine, you can now add 1 or 2 pounds sugar for each gallon of liquid.
7. When the liquid is lukewarm, add ¾ ounce yeast, dissolve in ½ cup warm water, with or without toast. Cover and stand in a warm place (65°-70°) to ferment for ten to twenty-one days.
8. At the end of the fermentation, strain, then rack. You will find this wine quite cloudy, so you will have to fine it with isinglass or eggshells.
9. When clear, bottle and cork.

The wine must be given a chance to mature (at least one year) before drinking.

II

Here is another recipe that will make a heavy, spiced wine.

You need: *6 lbs. pitted dates*
2 ¼ gals. water
4 lemons
4 oranges
1 oz. fresh ginger root*
12 cloves
2 lbs. sugar (optional)
1 handful date pits
¾ oz. yeast (3 packages)

1. Cut up the dates and put them into an enamel pot containing the water.
2. Wash and slice the lemons and oranges and add them.
3. Add the ginger root (bruise it first a little) and the cloves.
4. If you want the wine sweet, add the sugar and let it dissolve.
5. Bring slowly to a boil and let it simmer for one hour.

* You can also use about half the amount of powdered ginger, but I have found fresh ginger root preferable. It can usually be found in "foreign" sections of large cities and sometimes—as in California—just in ordinary markets.

6. Strain into a crock, put in the handful of date pits, and let the liquid cool to lukewarm.
7. Add the yeast, dissolved in ½ cup warm water.
8. Ferment and clear, as in the preceding recipe.

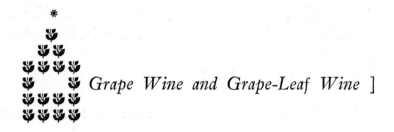 Grape Wine and Grape-Leaf Wine]

THE BLUENOSES SAID:
"There is a devil in every berry of the grape."
I say:
"There is joy in every berry of the grape."

I cannot hope to suggest homemade grape wines that will come anywhere near any good vintages. For the making of wine from grapes is one of the arts of the world, and it has been one of the pilgrim bearers of joy and exultation in the world from Noah to today.

There are endless kinds of wine of the grape made in all lands of the world, in vintages from palatial magnificence to humble peasant rough-reds. Wherefore it is hardly worth the effort to make grape wine in your own home. Yet I was enriched by a gift of a gallon of homemade wine by Officer Abelone of Danbury, Connecticut, made by his father. The latter is an Italian of the golden years when we were not worried by scientific progress and journalistic hullaballoo, when men were still individuals, not stamped out on assembly lines of TV and high-pressure advertising. The wine was made in the back yard of Mr. Abelone's home and had the scent and flavor of the pagan vineyards of Italy. It smelled and tasted Dionysian.

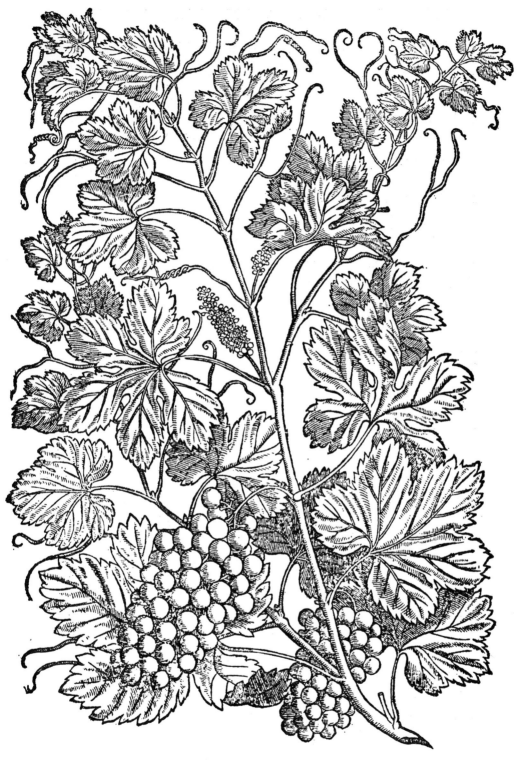

GRAPE

Nor shall I speak of the lore and the history and art of the grape. Books without end have been written about them. Wine and the vine are in divine paintings and in the dazzling white sculpture of the world. In pleasure and poetry they reign in amaranthine grandeur. They form one of the great joys in my daily life, as in the lives of many other men.

If you grow your own grapes, here are some very simple ways of making wine, if you wish to.

What I do recommend, however, is the making of the wine from young, tender grape leaves. This practically unknown beverage has a faint bouquet of grapes, plus an herblike aroma that makes a pleasurable gustatory and olfactory experience.

But first the recipes for grape wine.

The Recipes

I

You need: *any quantity of grapes*
some sugar, if you want a sweet wine

1. Take any amount of grapes, either grown or bought, put them in a crock, and crush them by hand. Let them stand for a week, well covered, for more fruit flies get into grape wine than into any other.
2. Fermentation will commence in a day or so. Grapes have their own yeast, which should develop by itself. If it doesn't, add ½ oz. yeast in the usual manner.
3. When the fermentation is finished, pour off the juice, squeezing the grapes in a press or through a thick cloth by hand. It is not easy work. Don't throw away the squeezed grapes (see following recipe).
4. If you want the wine sweet, put in 1 or 2 pounds of sugar for each gallon of wine.

5. Let the wine stand for a few more days, well covered. Then rack a few times and bottle.

II

It is possible to make an additional amount of wine from the residue of the grapes that you have squeezed. I have done it.

For every 4 pounds of leftover grape pulp, add 1 gallon water. Cover well and stand in a warm place. Fermentation soon begins. (Add yeast if it doesn't.) Let it continue for five or six days, then press the must through a cloth.

Add to every gallon of liquid 1 to 3 pounds of sugar in syrup form, depending on how sweet you like the wine. Let it stand in glass jars or gallons, not tightly closed, for a few weeks. I found a great deal of sediment at the bottom, and probably you will, too. Clear this by decanting.

Finally bottle. Do not drink the wine for at least two years. The longer you let it mature, the better it will be.

III

Here is the way you can make a white wine. Bear in mind that the juice of all grapes is clear; and it is the skin of the grape that gives the color.

Try to get Delaware or Catawba grapes, which contain a great deal of acid.

You need: *as many grapes as you wish*
1 gal. water for every 2 gals. juice
2 lbs. sugar for each gallon water

1. Mash the fruit and press the juice into a crock, either through a cloth by hand or through a press.
2. For every 2 gallons juice, add 1 gallon water in which you have dissolved 2 pounds sugar.
3. Let it ferment. (Add yeast if it doesn't.) When the fermentation

ends, clear into gallon jars, let stand for a few days, then rack into bottles.

Do not drink the wine for at least two years. The longer you keep it, the better it will be.

IV

Here is a way to make a grape wine in your kitchen in the city.

You need: *8 lbs. grapes (any kind)*
2 gals. warm water
4 lbs. sugar

1. Put the grapes into a crock. (You can halve all quantities in the recipe for a smaller amount of wine.) Mash them with a wooden spoon or squeeze by hand. Add the warm water. Add the sugar and dissolve it. You can add more sugar if you want a sweeter wine.
2. Let stand for about forty-eight hours, mashing the mixture three or four times a day.
3. Strain off the liquid into an enamel pot, squeezing out every drop of juice from the grapes through a cloth. Taste the must (which should have begun fermenting by then). If you want the wine sweeter, you can add more sugar at this point. If the must has not begun fermenting, which is very unlikely, heat it to lukewarm and pour it back into the crock. Now dissolve the yeast in ½ cup lukewarm water and put it in.
4. When fermentation has ceased, strain, fine, and decant.

Like all grape wines, this should not be drunk for at least two years.

All these young wines have the flavor and bouquet of fresh-squeezed grapes. Somehow they bring to mind the wine-drenched orgies of the bacchantes in the wild Greek woods. Some native-made commercial wines have the same bouquet and flavor. It is a classic flavor, hurtling through the twilight years when wine played an even greater and more intimate part in life than it does today.

Grape-leaf Wine

I have on my farm many wild grapevines, which of late have not borne any grapes, but the young leaves make a rare-tasting, delicate drink with a haunting tang—and easy to prepare. If you live in the city, you can always find some place in the country nearby where wild grapes grow in neglected fields, along fences, and up the trunks of trees.

You need: *8 lbs. young grape leaves*
           2½ gals. water
           4 to 5 lbs. sugar
           ¾ oz. yeast (3 packages)

1. Put the grape leaves into a crock.
2. Boil 2 gallons water and pour it over the leaves. Let stand two full days.
3. Pour off the liquid, squeezing every drop out of the leaves.
4. Pour an additional ½ gallon of very hot water on the squeezed leaves and spin them around with a wooden spoon for a few minutes, then squeeze every drop of water out of them. Add this to the first liquid in the crock.
5. Put in the sugar (5 pounds if you want a sweet wine) and let it dissolve.
6. Dissolve the yeast in ½ cup warm water and add it to the liquid.
7. Cover and set in a warm place (65°-70°) and for the first three days stir it once a day. At the end of the fermentation, strain off into glass jars or gallons, and if it does not clear by itself in a few days, fine with eggshells or isinglass. Then rack, bottle, and cork.

Do not drink this wine for two years. Then it will be a surprising pleasure.

Grapefruit Wine

IN MANY PARTS OF England grapefruit is sold under the name of "forbidden fruit" because it, too, was suggested as *the* forbidden fruit in the Garden of Eden. It makes a good, dry wine with a faint citrus tang. It is an ideal wine to make in the winter when the fruit is abundant.

The Recipes

I

Here is the simplest method.

You need: *14 grapefruit (good-sized)*
3 gals. water
3½ lbs. sugar for each gallon of water

1. Wash the grapefruit well. Peel as thinly as possible and put the rinds into the crock.
2. Squeeze the fruit by hand or with a squeezer and put the juice into the crock. There is no harm in having some lumps of fruit, but not the thick white part of the skin.
3. Boil the water in an enamel pot. Dissolve in it 3½ pounds sugar for each gallon of water. Pour onto the juice and rind.

4. This wine should ferment without any yeast. Cover it well, put it in a warm place (65°-70°), and let it stand for three days, by which time the fermentation should be well under way. If it is not, dissolve ½ ounce yeast in ½ cup warm water and pour it in.
5. Stir it twice a day during its active fermentation. When this is over, strain the wine into glass jars and let stand a week or two. By then the wine should be pale clear. If it is not, fine with isinglass or eggshells.
6. When clear, bottle and cork.

This matures quickly, and may be drunk after six months.

II

This recipe will give as good a wine, with a slightly different flavor.

You need: *1 lb. barley*
2¼ gals. water
10 grapefruit (good-sized)
8 to 10 lbs. sugar

1. Boil the barley and water for one hour.
2. While the barley water is boiling, wash the grapefruit, cut them into fairly thin slices, and put them into a crock.
3. Strain out the barley (you can use it in cooking, if you wish). Dissolve the sugar in the barley water. Vary the amount of sugar according to how sweet you want the wine to be. Pour the liquid over the fruit.
4. Cover and put in a warm place. The must should start fermenting without yeast. If it does not, proceed as I suggested in the first recipe.
5. At the end of the fermentation, strain and let the wine stand a week or two or even longer. If it does not clear perfectly, use

either isinglass or eggshells to clear. When all cloudiness has disappeared, bottle and cork.

This wine matures quickly and can be drunk after six months. It will have a bittery taste that is quite pleasant.

III

You can make an exciting grapefruit sparkling wine with a very piquant taste.

You need: *10 to 14 grapefruit*
2 gals. lukewarm water
7 lbs. sugar

1. Wash the grapefruit well, cut them into thin slices, and put them into a crock.
2. Pour over the fruit the lukewarm water and the sugar, stirring thoroughly. Fermentation should begin in a day or two; if it does not, you can help it with ½ ounce yeast.
3. Stir the fermenting must morning and night, removing all the bubbly scum that rises to the top.
4. When the fermentation is finished, strain and let stand in glass containers a week or two, then fine if necessary.
5. Bottle the wine in champagne bottles. Put a lump of sugar into each bottle. Stopper the bottles with champagne corks and wire cages.

The champagne wine will be ready to drink in two months. Don't keep it too long.

LEMON

 Lemon Wine]

THE COURTLY LEMON TREE, too, has the stigma of ranking among those supposed to have been the forbidden tree in the Garden of Eden. "The fruit of that forbidden tree whose mortal taste brought death into the world and all our woes," cried Milton. We are told: ". . . that even as the Palm was symbolic to Mohammedanism, the Passion Flower to Christianity, the citron or 'etrog,' the 'Godly Tree,' was the symbol of Judaism." To this day, this citrus fruit has an important part in the Hebrew Feast of Tabernacles. On mornings of that holiday, it is borne in the left hand into the synagogue; part of the ceremony is to smell its fragrant scent.

How mores differ among different peoples! In Bengal in India, and also in Ceylon, they say that the souls of vile ogres are gathered in the fruit. Then along comes a little boy who cuts the lemon into bits, and all the ogres die.

Of course, lemons have been, and still are, used in flavoring, for their scent, and for their medicinal values, but here is something special—for maidens only! If you want to learn whether or not you will win the man you love, carry a lemon peel in each pocket by day, and at night rub the four posts of your bed with them. If fortune favors you, the man will appear to you in your dreams and give you two lemons. This is gospel truth, as stated in *The True Fortune Teller*—a very valuable book in former days.

And so let us pass from loving ladies to good wine for you to make.

The Recipe

This is the simplest way to make a good lemon wine.

You need: *20 lemons*
 2 gals. water
 8 lbs. sugar
 1 lb. raisins
 1 slice toast
 ½ oz. yeast (2 packages)

1. Peel the rinds of ten lemons and put them into a crock. Peel the other ten lemons (you can dry these rinds and keep them for future use). With a sharp knife, cut away from all twenty lemons as much of the white part as possible. Then cut all the lemon pulp into small pieces and put them into the crock.
2. Boil the water for ten minutes and pour it into the crock. If you use pure spring water or water from a deep-sunk well, it need only be brought to the boiling point.
3. Cover and let stand for five days, stirring and mashing the lemon pulp once a day.
4. On the sixth day, strain the liquid into an enamel pot and add the sugar. Let this dissolve thoroughly while heating the liquid slightly.
5. Return to the crock and add the raisins, cut up slightly.
6. Prepare a slice of toast. Cover both sides with the yeast, dissolved in ½ cup warm water, and add both to the lukewarm liquid.
7. Cover and let stand in a warm place (65°-70°) for fourteen to twenty-one days, until the visible fermentation ceases.
8. Strain into glass jars and let stand for a week or two. Generally it will come golden clear of its own accord, but if it does not, fine, decant, and then bottle.

It should not be drunk for at least a year. Then it will be an excellent wine with fish, or drunk as a "spritzer"—adding one third charged water—on a hot afternoon.

Muskmelon (Cantaloupe) Wine]

THERE IS A FAIR and spacious garden on my farm filled with fruit and herbs—all thanks to her whom I chose among the daughters of the land as mine helpmeet. And she is a true helpmeet in that garden. There, as in other realms, I do all the talking and she does all the work; and naturally, I get all the credit. Let me add quickly that one of her great pleasures in life is to work among these fruits of the earth and scented herbs. She is avidly adventurous in this field. No sooner does the deluding, Easter-colored, siren-calling Spring Garden Catalogue arrive, than all atomic problems around us are forgotten and all life becomes one grand hope. Doesn't the catalogue show the most splendid, the latest variety of vegetables-to-be! So fruits from the South and the North are ordered and . . . strangely enough, some of them really survive climate and pests and rabbits and moles and deer. Thus we have had our own peanuts and yams and sweet potatoes and soybeans, every kind of herb and watermelon and muskmelon. One year the muskmelons, under my lady's eagle eye and constant toil, were especially excellent, and so a wine had to be made from them. And an excellent wine it was, and you will find it so, too, whether you make it from purchased or home-grown melons.

You may be interested in the history of the muskmelon, even as

I am. If you are not, just skip this. But if you are, read on and learn a little about your breakfast and dessert fruit.

It has a very ancient past. It was cultivated even before history was written. In Egypt, in Palestine, in all the Far Eastern lands it served as food, medicine, and drink, and was a very popular dish for the poor in those days. The seeds of the melon were salted, roasted, and eaten.

There was one Hebrew prophet, however, who did not like the melon: Elisha. And like all the flaming prophets of Israel, he was a great one for wild anger and fierce invectives. He cursed with violent curses the helpless melon. The result is seen on Mount Carmel in the Holy Land: a field of rounded, hard stones, once good-eating melons, but transformed into hard, inedible rocks by the wrath of Elisha.

The uncursed melons have continued fine food and make fine wine. Here is how it is done.

You need: *12 lbs. muskmelons (after rinds and seeds are discarded)*
2¼ gals. water
5 lbs. sugar
2 lemons
½ oz. yeast

1. Buy (or grow) about 15 pounds ripe muskmelons and pare the rinds quite thickly (leaving 12 pounds of fruit). Cut the melons into small pieces and put them into a crock.
2. Boil the water, cool it a little, and pour it over the cut fruit.
3. Mash it with your hands or with a wooden spoon and let the mash stand for twenty-four hours, stirring it often with a wooden spoon.
4. Strain the liquid into an enamel vessel, clean the crock, and pour the liquid back into the crock.
5. Dissolve the sugar thoroughly in the liquid.

6. Wash the lemons. Cut the rinds very thinly and add them, together with the squeezed juice.
7. Pour a good quantity of the liquid—about half—into the enamel pot and heat it well, then pour it back into the crock. This should warm it to lukewarm, just right for the yeast.
8. Dissolve the yeast in ½ cup warm water and add it to the liquid. Let it ferment for about ten to twenty-one days, stirring it once a day during that period.
9. When the visible, active fermentation is over, strain the wine into gallon jugs or gallon jars. Let it rest for a few days, and if it does not clear completely, fine and rack until it is perfectly clear. Then bottle and cork.

You can make wines from other melons in the same way. Watermelon wine was quite common around New Orleans in the olden days. Honeydew melon should also make a flavorsome wine.

 Orange Wine]

I HAVE FOUND ORANGE WINE soft and pleasant, perfumed like the sacerdotal blossoms and fruit.

On a May-filled sunny day, the journeying Holy Family—the blue-clad Virgin Mary, her radiant infant, Jesus, and sturdy Joseph—came to a blossoming orange tree. Before it stood a fierce beaked eagle, guarding it.

"An orange for my sweet Son!" the Holy Mother pleaded of the bird.

But the task of the royal bird was to guard, not to give, and he refused.

Mary looked at him with her softly glowing eyes, and the bird fell sound asleep.

She plucked three oranges, one for her spouse, Joseph, one for her wondrous Son, and one for her humble self. When she had done this, the eagle awoke. He saw each one of the Holy Family holding a golden-colored fruit; but he knew that it was not for him to take the fruit from them, and so he let them go forth, shining fruits in hand.

It is a tale befitting the "apples of gold." To this day, in Sicily, statues of the Madonna are decorated with branches of orange blossoms, and in some towns, posts are set up decorated with orange boughs.

Oranges, like so many other fruits, are designated as *the* forbidden fruit eaten by Eve. It is interesting to note that very many tragedies in the world have been caused by women, beginning with the commencement of time. That is most gratifying. It is always such a pleasure to blame, conveniently, others for our woes. While on the subject of women, oranges and orange blossoms have had, and still have, an important place in their lives. The orange tree is one of those that bears leaves, flowers, and fruit at the same time. Hence, in some lands, it is the symbol of perfect fructifying. Beginning with the Saracens, orange-blossom wreaths have been worn by brides as an emblem of a full and happy marriage. The custom is still followed today, without any thought of its folkloristic meaning. Probably this particular symbolism commenced even before the Saracens. Juno gave to Jupiter on the day of their marriage a "golden apple" —surely an orange—and he probably gave her orange blossoms in return. Perhaps it was a marriage rite among the gods. These golden fruit grew only in the garden of the Hesperides, near a dragon who slept most of the time. It was Hercules who succeeded in carrying off some of them.

"Golden apples" may have been golden oranges, given by Aphrodite to Hippomenes to win the famous race against Atalanta.

And coming down to modern days, the good folks of Portugal claim that the original tree brought from China to Europe, from which all European orange trees have sprung, is preserved in Lisbon in the garden of the Count of Saint Laurent.

Yet, to dream of oranges always presages misfortune. So, to counteract this, we will turn to good orange wine, which always brings good spirits.

The Recipes

I

Part of this very simple recipe was given to me by Bertha Nathan, of Maryland.

You need for this only sugar, oranges, toast, yeast, and water.

1. Squeeze 4 quarts orange juice and pour it into a crock.
2. Place 1 gallon water and 2 to 4 pounds sugar in an enamel pot. Boil for 20 minutes.
3. Pour this into the crock, and also add the thin rinds of five well-washed oranges.
4. When lukewarm, prepare a slice of toast and ½ ounce yeast, dissolved in ½ cup warm water, and put them into the crock.
5. Let the must ferment, then strain, clear, fine, and bottle.

If you don't put in much sugar, you will have a nice, dry orange-bouquet wine.

II

It is best to use the juice of about twelve to fourteen medium-sized oranges for each gallon of water. If you want a greater quantity of wine, multiply each unit accordingly. I'll assume that you will make 2 gallons of wine.

You need: *25 oranges (medium-sized)*
1½ gals. water
4 lbs. sugar
rinds of 5 oranges
1 slice toast
½ oz. yeast (2 packages)

1. Squeeze the juice of the oranges into a crock.
2. Put the water, sugar, and orange rinds into an enamel pot. Bring this slowly to a boil and let it simmer for twenty-five minutes, removing any impurities that rise to the top.
3. Strain, pour into the crock of juice, and let cool to lukewarm.
4. Cover a piece of toast with the yeast, dissolved in ½ cup warm water, and put them into the crock. Cover and let ferment—about fourteen days or more.
5. Strain the wine into jars and let it rest. If not clear after a week or two, fine and finally bottle.

If you don't drink this for a year or more, you will have an excellent and unusual sauterne-like wine.

III

Here is an ancient recipe that I tried, and I think you should, too. It makes a magnificent wine.

"To six gallons of Spring-Water put twelve pounds of single-refin'd Sugar, the Whites of four Eggs well beaten, put these to the Water cold; then let it boil three-quarters of an Hour, taking off the Scum as fast as it rises: When 'tis cold, put in six spoonfuls of Yeast, and six ounces of Syrrup of Lemon [I made mine from lemon juice and sugar], beaten together; put in also the Juice and Rind of fifty large Oranges thin pared, that no White-part nor any of the Seeds go in with the Juice, which should be strained: Let all this stand two Nights and two Days in an open Vessel, or large Pan, then put it into your close Vessel, and in three or four Days stop it down: When it has stood three Weeks thus, draw it off into another Vessel, and add two quarts of Rhenish or White-wine; then stop it close again, and in a Month or Six Weeks 'twill be fine enough to Bottle, and to drink in a Month after. If you wish to keep it, put in Brandy instead of Rhenish."

It sounds complicated, but it is not. Halve the amount of each ingredient, and you will have a comfortable amount for home wine making.

[Peach Wine]

IT IS INTERESTING to note that the Chinese poets, in mind and word, have been more cognizant then any other peoples of the loveliness of the peach blossom and its fruit.

If you have a peach tree or know of one in a friend's garden, look at the delicate-colored and delicate-scented blossoms on warm spring days, and then, later, at the young fruit, and finally at the fully ripened fruit, and you will see a poem of perfect harmony. The rare, roseate delicacy of its dreamy coloring and its fragile fragrance bring to mind the sun, air, wind, and dew.

> "The young peach with its fresh color
>   is beautiful,
> Its reflection on the ground scatters
>   fragrance"

sang Emperor Chien Men of the Liang dynasty.

The color of the fruit fairly vies with the color of pink clouds in the early morn, and there is an airiness about its scent that cannot be told in words.

The peach has lived fully in China in symbolism, art, tales, and poems. The blossom is a symbol of matrimony and a herald of spring. The fruit is an emblem of long life. There is a giant peach tree in China that blossoms once every three thousand years. It belonged to the Mother Queen of the West Heaven, Hsi Wang Mu. On her birthday the immortals assemble at the Feast of Peaches to celebrate the occasion. Since I made my first peach wine and peach brandy, I have a Feast of Peaches every time my friends and I drink them.

Brooms of peach branches are used in China to sweep away witches, and a distillation is made of the flowers to ward off evil. Under "mulberry wine" you will find a recipe using ashes of the mulberry tree and peach-tree gum to prepare a drink that eliminates the necessity of food and illumines your body.

There is a charming Japanese fairy tale dealing with the peach. Once when an old woman was washing her clothes in the river she suddenly saw the water splashing and rolling as in a great storm. But it was not caused by any storm, but by a tremendously big pink-and-white round object, rolling and tossing under the water. She pulled it out. It was a beautiful giant peach, enough for her and her husband to eat for weeks to come.

She rolled it to the bank and opened it to taste the fruit, and there, in the center, instead of a pit was a beautiful child, a boy with cheeks as pink as the fruit. She brought the child to her house and she and her husband were very, very happy, for they had no children and had wanted one for many years.

The boy grew up strong as a giant, and they sent him to school. He was as brilliant in school as he was strong.

Then came the day when devils invaded the island, but the Peach Boy destroyed them and acquired their treasure, which he gave to his parents as a reward for the love and care they had given him.

A fine Alger hero tale, showing life as it should be.

In medieval times peach kernels were often used in cosmetics. And almost to the present day there was the belief in Sicily that any sufferer from a goiter would lose it by eating a peach on Ascension Day—if the peach tree died at the same time. In other parts of Italy, warts are charmed away when peach leaves are buried in the earth.

After burial—even for warts—a good drink is in order! Here is how the good wine is made.

The Recipes

I

You need: *12 to 15 lbs. ripe peaches*
6 to 8 lbs. sugar
2 gals. water
2 lemons
2 oranges
¾ oz. yeast (3 packages)

1. Use freestone peaches if possible, though this is not essential. Cut them into halves or quarters and put them into a crock. Break up a third of the pits and add them, together with the kernels. Add a gallon of water, mash the fruit with a wooden spoon, cover, and let stand for forty-eight hours.
2. Now dissolve the sugar in a gallon of water, heat it well, and pour it into the crock. (The amount of sugar you use is determined by how sweet you want the wine to be.)
3. Cut the rinds of the well-washed lemons and oranges thinly (you can substitute additional lemons for the oranges). Add them, as well as the juice of the fruit. Cool (or warm) the liquid to lukewarm.
4. Dissolve the yeast in ½ cup of warm water and pour it into the

PEACH

PEAR

crock. Cover and let stand while fermentation is in full swing. When it has subsided the least bit, strain off the pulp and rinds and clean the crock, then put the fermenting wine back into it. Wait until fermentation has fully stopped, which may take from ten to twenty-one days, strain the wine into glass jars, and let it rest for a week or two. If not perfectly cleared, fine with eggshells or isinglass. Then bottle and cork.

Do not drink the wine for six months to a year.

II

You need: *40 to 50 peaches*
2 to 4 lbs. sugar
2½ gals. water
2 lemons
2 oranges
¾ oz. yeast (3 packages)

1. Cut up the peaches (taking out the stones and putting them aside for later use), and put first a layer of peaches on the bottom of the crock, sprinkle it with a layer of sugar, then add another layer of peaches and another layer of sugar, repeating this until all the sugar is used up.
2. Cover and let stand for twenty-four hours.
3. Boil the water in an enamel pot, together with additional sugar if you want a sweeter wine.
4. While the syrup is boiling, add the sugared peaches and their juice and continue cooking until the fruit becomes mushy. Clear away any impurities that come to the top.
5. During this time, break open about half the peach pits and throw them in.
6. Pour into a crock and add the washed, thinly pared rinds of the lemons and oranges (you can substitute lemons for the oranges) and their juice.

7. When the liquid is lukewarm, dissolve the yeast in ½ cup of warm water and put it in. Cover and let stand. The fermentation will begin soon, and you let it continue for one week.
8. At the end of that time, strain the must through a thick cloth into an enamel pot, clean the crock, and pour back the must. Now wait until all fermentation has ended.
9. Strain the wine into glass jars and let it rest for two weeks or so. If it does not clear, fine; then bottle and cork.

This wine should not be drunk for at least a year.

 Pear Wine]

YES, pears will make a pleasant, light, dry wine, almost like a Graves. It is a delightful drink on a hot summer's afternoon, and can be made in the city as well as in the country.

Of course, the fruit and the tree have a living lore of their own, which for me adds a greater pleasure to the wine. It is good to taste a wine with your mind, even as you taste it with your tongue.

The pear was consecrated to Venus, and there was a well-known "pear of love." The Gaelic legend transforms the mystic apples of the Isle of the Blessed to a species of pear. In many parts of Europe there is the romantic and practical custom of planting a tree for a newborn child—a pear for a girl, an apple for a boy—for in the dim years these trees were thought to have magic powers to repel evil spirits. Thus, the Circassians plant a pear tree to protect their cattle, and often keep one in the house as a sort of minor divinity. Many honors are paid to it with singing and dancing. If you dream of ripe pears, good luck is coming your way; if the pears are unripe, it means the opposite. And girls, except princesses, particularly like to dream of ripe pears, for this means marrying far above their station.

And with that, to the wine!

The Recipe

You need: *10 to 15 lbs. pears (any kind, so long as they are ripe and juicy)*
2 gals. water
3 to 4 lbs. sugar
4 lemons
2 oranges
1 lb. raisins
¾ oz. yeast (3 packages)

1. Wash the pears, cut them into small pieces, and put them into a crock.
2. Add to the water 3 to 4 pounds sugar (depending on how sweet you want the wine). Bring to a boil, then pour the syrup over the pears and mash them with a wooden spoon as thoroughly as possible.
3. Add the thin rinds and juice of the lemons and oranges, and the raisins, cut up a little.
4. Dissolve the yeast in ½ cup warm water; add it to the liquid when it is lukewarm.
5. Cover, set in a warm place, and let it ferment, stirring once a day.
6. When the fermentation has ended, squeeze through a cloth and strain into glass jars.
7. Let the wine stand for a week or more to see if it will clear. If it does not, fine and then bottle.

Don't drink the wine for at least one year. A cider can also be made from pears, just as from apples.

[*Pineapple Wine*]

"FRUIT OF ALL FRUITS the first, which are, which were, which will be, Pineapple most eminent of fruits." So the pineapple was hailed by the poet Lochner.

The fruit is a native of tropical America and the West Indies and was not known until the new continents were discovered. Its native name was *na-na*, meaning "fragrance." Hence the French name, *anana*, and the Portuguese, *ananaz*. The Spaniards called it *piña de Indias* because it resembled a pine cone, and the English name has the same derivation.

The praises of this rich fruit have been sung from the time it was first discovered and eaten. As early as 1555, Jean de Lery, a Protestant minister who came to preach on the new continent, speaks of the pineapple as "worthy of the Gods and of such excellence that it should be picked only by the hands of Venus." It requires a Venus with strong hands for the purpose. If you have eaten the fruit where it is grown and ripened, you will agree with the good cleric.

Charles de Rochfort, writing a *History of . . . the Caribby-Islands* in about 1658, says in part:

"The Ananas or Pine-Apple is accounted the most delicious fruit, not only in these Islands, but of all America. It is so delightful to the eye, and of so sweet a scent, that Nature may be said to have been extramly prodigal of what was most rare and precious in her Treasury to this Plant." (This from the English translation by John Davies, published in London in 1666.)

For the next two hundred years, the opinion never changed. Sebastiāno Rocha Pitta wrote in 1730:

". . . the first of these is a pineapple which is the King of them all, Nature has crowned with a diadem of its own leaves. The latter surround it with thorns and guard it like archers."

I don't know of any fruit that has been used as much in early American art designs as this—in architecture, furniture, silverware, gateposts, carvings over doors and on hitching posts as a sign of hospitality, and on bedposts for the same reason.

I am writing this chapter on a beautiful dark mahogany desk more than a century old, with strong legs carved in pineapple designs. At the gate into our farm there is an old iron hitching post crowned with an iron pineapple—a sign of welcome and hospitality.

I believe the vogue for carving this fruit began with sailors bent on long voyages, who preferred to carve live, exotic Dominican mahogany rather than cold ivory or yellowed bone, and preferred realistic nature to looming fables. Then, in port, folks liked these heavy carvings and bought them, and the tropical fad spread far and wide.

It is nice to write of a tropical fruit and a tropical wine today—in the most untropical scene conceivable. It is the 21st of March. Wide, thick snowflakes are coming slowly down from the gray heaven through the yellow-gray air on this first day of spring. They are falling on the bare branches of the trees, forming exquisite patterns of silvery flowers. On the snow-decked, gently sloping hill, a big shaggy black dog is circling, bouncing and leaping around, dancing his ecstatic spring rites to the Snow God, who must now leave the land.

Every now and then I stop writing to fill my mind and eyes with this silent-wild music of nature. Then I get up and turn to what I have been doing all morning, intermittently with this writing—making pineapple wine and pineapple brandy. This is how to make the wine. The brandy will be found in the section on brandies.

The Recipes

I

You need: *peeled, cut-up pineapple fruit to fill 8 quart jars*
2 gals. water
5 to 8 lbs. sugar
¾ oz. yeast (3 packages)

1. Take ripe pineapples, peel them, and cut them into pieces. If you want only one gallon of wine, use half the quantities. Put the cut-up pineapple into a crock.
2. Boil the water and pour it over the fruit in the crock. Cover and let stand for four days, stirring every day.
3. Strain off the juice into an enamel vessel, squeezing every drop out of the fruit through a cloth.
4. Clean the crock and pour back about half of the liquid. Into the remaining juice put 5 pounds sugar (more if you want it sweeter). Dissolve thoroughly and heat the liquid quite hot. Pour this back into the crock, thus giving you a lukewarm liquid.
5. Dissolve the yeast in ½ cup warm water and add it.
6. Cover and set in a warm place (65°-70°) to ferment for ten days. Sometimes it will take a little longer. Siphon off the top clear liquid and strain the rest through a thick cloth into glass jars.
7. Let stand for two weeks. If not clear by then, fine with eggshells or isinglass. If the top part of the wine is clear, siphon it off first. When perfectly clear, decant and bottle.

Don't drink it for one year. It is pleasantly dry, with a pineapple flavor and bouquet.

II

If you have a small press, you can make your pineapple wine differently. Don't try this without a press.

You need: 2 gals. water
5 to 8 lbs. sugar
8 to 10 pineapples
2 lemons
¾ oz. yeast (3 packages)

1. Boil the water, into which you have put about 6 pounds sugar, and pour it into a crock.
2. Peel the pineapples, cut them into chunks, and put them into the press. When the juice is all pressed out, pour it into the crock.
3. Pare the lemons thinly. Place the rinds and juice in the crock.
4. Dissolve the yeast in ½ cup lukewarm water and add it. Cover and let ferment until active fermentation ends.
5. Strain the wine into glass jars and let it stand for two weeks. If not completely clear, fine with eggshells or isinglass, and then bottle.

This wine also needs at least a year's maturing.

III

In olden days, folks in Louisiana made a pineapple wine called "bierre douce" (sweet beer). It was made from pineapple peelings, brown sugar, cloves, and rice; it was highly spiced and very strong. It can easily be made now, following the instructions of the previous recipes, and is well worth trying.

Substitute the thick peelings of ten pineapples for the pineapple fruit, and put them in the cold water together with a dozen cloves and 2 pounds of rice. Let this boil lightly for one hour. When it has cooled to lukewarm, strain into a crock and add 6 to 8 pounds brown sugar (dissolving it thoroughly) and the rind and juice of two lemons. Finally put in ¾ ounce yeast (3 packages), dissolved in warm water. Then continue as in the previous recipe.

Plum Wine]

IF YOU DO NOT GROW plums in your own garden, buy them from an orchard or a fruit store when they are in season. They will make an excellent wine or brandy with the delicate flavor of a commercial Yugoslavian slivovitz or a good port. Some of the flavors of these fruit and flower wines are difficult to describe. All you can say is that they are different from any wine flavors you have ever known.

Boughs of plum trees on the doors and windows keep witches away in some parts of Europe, yet to dream of plums means tragic trouble for the dreamer. It means infidelity, if you are in love or married. It means that ill health is on the way. It means a loss of wealth. So just avoid dreaming of plums! But in China the plum is a symbol of friendship. It is a favorite tree of that land of ancient wisdom: it has been praised in poetry, music, and literature, and by custom.

Chin-Ling-Chih, a princess in the Sung dynasty, one day fell asleep under the eaves of the Hang-Chang hall of the palace. A plum blossom fell on her white forehead and left on her royal brow the mark of its five petals in delicate cream rose, which neither hand nor water could remove. Thenceforth all the maidens of the palace copied the princely mark on their foreheads.

In that multipeopled land there are many more tales of the plum fruit, the blossom, and the tree. Here is one charming to me and . . . to bachelors.

In the *Sung Shih* there is the biography of a hermit, Lin P'u, who desired the life of a peaceful hermitage in preference to cacoph-

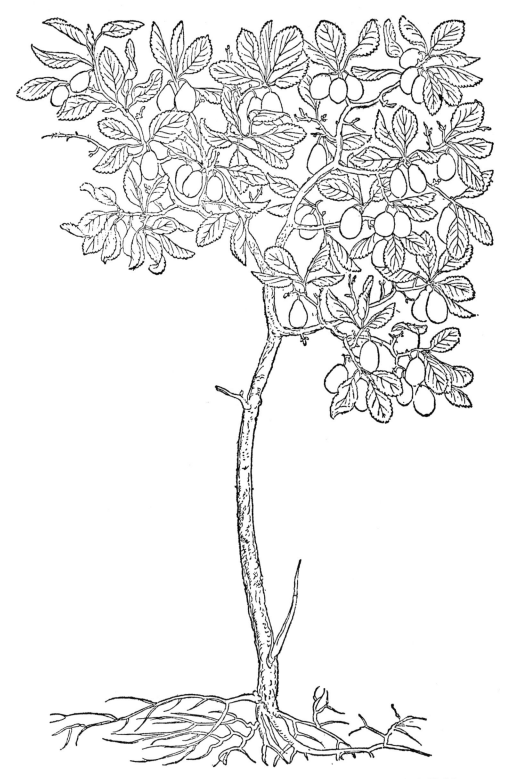

PLUM

onous cities and peoples. To secure perfect peace, he had neither wife nor children. He built himself a hut on the Ku Mountain, near one of the lakes, planted a plum tree, and gathered around him soft-colored cranes. In this paradise on earth he called the sturdy plum tree his wife, and the rose-and-white slender-legged cranes his children. . . .

Both in China and Japan the plum tree is a symbol of longevity and immortality: the contrast of its knotty trunk and new green shoots shows that despite age, youth can rise anew, therefore can hope. Its companion is the nightingale, the joyous harbinger of spring, since the plum is the first of the flowering trees to emerge from its wintry mantle.

What is lovelier than this spring flow of words by Yuan Huai, of the Yuan dynasty?

". . . This morning the Plum's rich color looks like the apricot
Opening in the lone village beyond the small bridge.
Perhaps the creator does not like pale shades
So invented it to add to its charms."

The plum also had an important place in the culinary art of China. Here is a delicacy worth trying. It is a recipe for plum-flower dumplings from the *Sha Chia Ch'ing Kung:*

The flowers of white plum and sandalwood are powdered and mixed with water, and when the water has absorbed the fragrance, the powder is removed. Flour is then mixed with the water to make a dough. Of this, thin flat dumplings in the shape of five-petaled flowers are cut and dropped into chicken broth. They should add a delicate fragrance to the broth. But the fragrance of the wine is much stronger.

The Recipes

I shall begin with "weina palama." This is a recipe that was sent to me by . . . alas! I lost the record; from . . . again alas! I don't know from where. I regret this and apologize humbly, for I would like to give proper credit to the sender. The recipe is very simple and easy, and I assure you it will make an excellent drink. Here it is:

I
Weina palama

"Make a mash of plums and sugar and water, to the proportion of 1 cup sugar to ½ gallon plums and just enough water to cover. Let this ferment for twenty-two days and strain.

"The plum used is a Spanish strain about the size, shape, and firmness of ripe black olives marked commercially as medium-sized.

"Once a group of coeds, housed in a church boarding home for women, had some anxious days because it was very difficult to mask the odor of the fermenting mash, especially in the second week of the process. The unmentioned history of the coeds and the wine leaves a great deal to the imagination and to conjecture."

I made the wine exactly in the same way, but I added a little more sugar. The wine became more delectable and potent the longer I kept it.

II

You need: *2 gals. water*
6 to 8 lbs. sugar
20 lbs. plums (any kind you like)
¾ oz. yeast (3 packages)

1. Put the water into an enamel pot and add 6 pounds sugar (more if you want a sweeter wine). Heat this to boiling.
2. Put in the opened plums, leaving in half the pits, and mash them thoroughly. Simmer gently for thirty to forty minutes, removing any scum that comes to the top.
3. Strain into a crock, squeezing all the juice from the plums, and let the liquid stand until it is lukewarm.
4. Dissolve the yeast in ½ cup warm water and add it. Cover and

set in a warm place (65°-70°) until fermentation has stopped.
5. Strain into glass jars and let stand for two weeks or more. If it is not absolutely clear by then, fine with eggshells or isinglass. Of course, you can take off any part that is clear and bottle it. When all the wine is clear, bottle, and do not succumb to the temptation of drinking it until a year has passed. It will then be good wine—and the longer it is allowed to mature, the better it will be.

III

You can make a very good spiced portlike wine of plums, as follows:

You need: *15 lbs. plums*
2 lemons
2 oranges
½ oz. ginger root
10 cloves
3 peppercorns
cinnamon (or any other spices)
2 gals. water
6 to 7 lbs. brown sugar
¾ oz. yeast (3 packages)

1. Open the plums, pit them, and put them into a crock.
2. Take a dozen of the pits, open them, and throw the pits and kernels into the crock.
3. Wash and slice the lemons and oranges and put them in, skin and pulp.
4. Make a spice bag of the bruised fresh ginger root, cloves, peppercorns, and cinnamon (or any other spices you favor). Suspend the bag by a string to reach deep into the crock.
5. Bring the water to a boil, pour it over the fruit, and stir vigorously.
6. Cover and let stand in a warm place (65°-70°) for a week, stirring it each day a few times with a wooden spoon.

7. On the eighth day, strain into an enamel vessel and dissolve in it 6 to 7 pounds brown sugar.
8. Warm the liquid and pour it back into the crock. Now dissolve the yeast in a little warm water and pour it into the crock.
9. Cover and stand in a warm place until fermentation has ended. Then strain the wine into glass jars. Let it rest for two weeks, and if not clear by then, fine, then bottle.

This will give you an excellent spiced portlike wine, good enough to warm any man on a cold day. Don't drink it for one year.

Pomegranate Wine]

"I WOULD CAUSE THEE to drink of spiced wine of the juice of my pomegranate." Thus the Song of Songs sings of the pomegranate in the great book of the world, showing that a wine was made from it in Biblical days. The very name of the fruit has a Far Eastern echo, and its history is rich in rose-colored poetry and tales. Let me tell you a few of its traditional gorgets of gilded legends and connotations. They are wordy fabrics of fantasy, rich as the taste of the fruit and the color of the wine.

It begins with the beginning of the world when, some say, this was the marvelous tree of knowledge in the Garden of Eden; and each fruit is said to contain one seed from the Garden. References to it ever run through prosals and thoughts: "Thy plants are an orchard of pomegranates, with pleasant fruits; camphire, with spikenard . . . and saffron."

In another land there is a dark tale of how the tree came to be —a tale of unrequited, illicit love that made the gods turn the loved one into the tree and the lover into a sparrow hawk. Ever since, no sparrow hawk ever alights on a pomegranate, but flies from it.

The culture of the fruit began in prehistoric days, and in addition to its use for wine and food, the flower served as a pattern for golden ornaments. They were embroidered and embossed on the robes worn by the Israelites in their temples. The calyx lobes of the fruit inspired the designs on Solomon's crown and all crowns thereafter, as well as the decorative ornaments of Solomon's temple.

In those days the fruit was not only eaten, but was the basis for spiced wine and other drinks. It was also used in medicine and for dyeing hair and cloth, not only among the Israelites but among other people of the East as well. The Persians spoke of it and its crimson beauty in song and poetry. The Egyptians held it sacred. In Turkey, as in many other lands, it was a symbol of fertility because of its many seeds. To this day you will sometimes find Turkish women throwing a pomegranate on the ground to see how many seeds will fall out of the broken fruit. That tells how many children they will have.

The Chinese, in particular, have innumerable songs of the pomegranate, for there the fruit is common everywhere.

> "The pomegranate opens,
>   A hundred sons."

One Chinese poet called it "the eighteenth aunt of the wind." Other poets compared the fruit to the smile of a beautiful girl because of the brilliant scarlet flames in the summer and shining green and red fruit during autumn days.

> "Their red beauty surpasses the evening glow.
> No painter can paint anything more beautiful."

So sang Liu K'o Chuang, of the Sung dynasty.

The pomegranate, even as the peach, was a symbol of good luck, long life, and many sons.

The Greeks spun around this fruit a wondrously beautiful tale. There was a nymph, delicate as gossamer in the dewy morning sun. A fortuneteller foretold that someday she would wear a crown.

The god Bacchus loved her, but never married her to put the golden crown on her head. Instead, he changed her into a pomegranate tree and twisted the calyx of the blossom into a crown. So she did receive her sign of sovereignty and has worn it ever since.

Persephone paid a heavy price for having tasted the fruit. When she was in the underworld with Pluto, Demeter, her mother, wanted her on Olympus among the gods, and Jupiter promised that if she did not eat any fruit of the underworld, she could return. But Persephone had eaten a lush pomegranate that she had plucked in the Elysian fields, and so she could not leave. Demeter, in wild fury, prevented all fruit from growing on the earth. Then Jupiter conceded that Persephone could go to Olympus for six months each year. It has been thus even since. And hence the fruit became a symbol of the blurred, dreamy, dark realm.

In those days the juice of the fruit was white; it was the blood of Menoeceus that turned it red, for he ended his life under the tree. This added to the dark symbolism of the pomegranate.

Let me turn to the sunny side of legend.

The great god Attis was born when his mother, a virgin, conceived him by placing a scented pomegranate in her bosom. The Greeks knew the scent well, for they made a popular perfume from the rind.

The fruit also played an important part in Christianity and Christian art. It was the symbol of hope, richness, faith, and immortality.

Saint Catherine holds a pomegranate in her hand, and the infant Jesus often holds the fruit in His hand, offering it to His Virgin Mother.

I have only touched on the lore that grew up around this incarnadine fruit. Through the centuries its presence and praises have ornamented every land, and now I will tell you how to make a wine from it that will ornament the hours when it is drunk. Make it, no matter what the cost of the fruit may be!

The Recipe

You need: *6 ripe pomegranates*
3 lemons
2 gals. water
4 to 6 lbs. sugar
1 slice toast
½ oz. yeast (2 packages)

1. Take the seeds from ripe pomegranates and put them into a crock, crushing them with your hands or a wooden spoon. Pare the lemons thinly and add rinds and juice to the crock.
2. Boil the water in an enamel pot into which you have put 4 pounds sugar (6 pounds if you want a sweet wine). Pour the boiling syrup into the crock.
3. Prepare a slice of toast and cover it with the yeast, dissolved in ½ cup warm water. Float the toast in the crock.
4. Cover and set in a warm place (65°-70°) to ferment. When the visible fermentation has stopped, strain through a heavy kitchen towel (I found this best for the purpose) into large glass jars.
5. Let the wine rest for two weeks or even more, to see if it will clear of its own accord. It should be a brilliant rose color. If it is not completely cleared at the end of that time, siphon off the top —if not cloudy—and fine the rest with eggshells or isinglass. When brilliantly clear, rack, siphon off, and bottle.

This will give a pleasant, tart wine. Like most, it improves with time and should not be drunk for at least a year.

 Prune Wine]

PRUNES MAKE AN EXCELLENT WINE. I have made it, and my son, too, has made it, and a fine time was had by all, including foreign graduate students at Johns Hopkins University.

As with all folk wines, there are many ways of making this one, for remember, folks vary, and each one has his or her own definite pernicketiness as to how this or that should be done, or how "my forebears" did it. Here are two recipes I have used.

I

You need: *5 to 8 lbs. sweet prunes*
2 gals. water
6 to 8 lbs. sugar
1 slice toast
½ oz. yeast (2 packages)

1. Wash the prunes and put them into a crock.
2. Pour the boiling water on them and let them stand, covered, for twenty-four hours, mashing them with a wooden spoon a few times during the day.
3. Strain them into an enamel pot, squeezing every drop of juice out of the prunes. Leave in about a handful of the pits, some cracked open.
4. Heat the liquid and pour it back into the crock.
5. Dissolve the sugar in it—the amount depending on how sweet you want the wine to be.

6. Prepare a slice of toast. Dissolve the yeast in ½ cup warm water, spread it over the toast, and float it in the crock. Cover and let stand in a warm place (65°-70°) until fermentation has stopped. This will take about two weeks.
7. Strain the wine into glass jars and let it stand for at least two weeks. If it is not clear enough to bottle, fine, and finally decant and bottle. Just remember, after the wine has rested for two weeks, to siphon off any part that is clear before fining the rest.

II

This is another good prune wine.

You need: 5 to 7 pounds dried sweet prunes
2 gals. water
6 to 8 lbs. sugar
2 lbs. raisins
3 lemons
4 oranges
½ oz. fresh ginger root
½ oz. yeast (2 packages)

1. Wash the prunes and put them into a crock containing the water. Cover and let stand for two weeks, mashing the fruit each day with a wooden spoon.
2. At the end of that time, strain into an enamel pot, squeezing every drop of juice out of the prunes. Heat the liquid and pour it back into the crock.
3. Put into the liquid 6 to 8 pounds sugar (depending on the sweetness desired) and dissolve it well.
4. Now add the raisins, cutting them up a little; the washed, thin rinds and juice of the lemons and oranges; the bruised ginger root; a handful of cracked prune pits; and the yeast, dissolved in ½ cup warm water.
5. Cover well and let stand in a warm place (65°-70°) from two to three weeks. If the wine is not sharply clear, fine, then bottle.

In one year you will have an unusually fine, sweet dessert wine equal to a good port or cream sherry. Two years later it will be better. The longer you keep it, the better it will be.

[*Quince Wine*]

THE QUINCE WAS CONSECRATED to Venus, and when a young Grecian man or woman gave this fruit as a gift to one of the opposite sex, it was like giving a wrist watch today—an offer and acceptance of a coming marriage. A newly married bride in Greece had to eat a quince before the marriage night. And when the Goddess of Love came riding along, her chariot was filled with roses, myrtle, and—quinces.

There is a charming ancient tale of how handsome Acontius won his bride with the aid of a quince. One holiday he came to Delos for the sacrifices to Diana, and there, for the same reason, had come Cydippe. She was lovelier than a flower in the morning breeze, and Acontius fell in love with her the moment he saw her. She looked at him, too, and blushed, and thought him handsome as a statue of Dionysus. But alas! Acontius was an unknown lad and Cydippe came from an honored and rich Athenian family. He could not ask so noble a maiden in marriage.

Acontius was quick of mind, even as he was handsome to behold. "Why not tell her of my love with the fruit of love, and win her at the same time!" he thought.

So he plucked a quince and inscribed on it: "I swear, by the goddess Diana, to become the wife of Acontius." Then he waited until he saw Cydippe and her friends go into the temple to pray. Acontius, unseen by anyone, then threw the quince into the temple.

It rolled before Cydippe, she picked it up and, with a girl's curiosity, read aloud to her friends what was inscribed on it: "I swear, by the goddess Diana, to become the wife of Acontius." Thus she had inadvertently sworn before Diana an oath she was bound to keep. Needless to say, she was overjoyed to do so. Her parents now were forced to give their consent to the marriage.

The Greeks called the quince (as well as the orange) the "golden apple," and thus the tales dealing with golden apples were associated with it: the golden apples of the Hesperides; the golden apples of Hippomenes, that won him Atalanta in that famous race.

On the practical side, the Romans, and even the Greeks, used the quince as a hair dye. The blossoms were made into a famous perfume in the city of Cos. Medieval necromancers used it in beauty concoctions.

In the Bible, the quince was yet another of the supposed fruits of the tree of knowledge. But the wine you will make of it will bring happiness instead of evil. Even dreaming of quinces foretells good luck, and in case you are in the company of anyone who has accidentally swallowed a deadly poison, the juice of a raw quince is a fine antidote. So said Nicholas Culpeper in *The English Physitian*. And so, to the wine.

The Recipe

You need many quinces, but they are inexpensive as a rule, and if you have a quince tree in your garden, it will give you all the quinces you can use.

You need: *15 to 20 quinces*
 1 ½ gals. water

QUINCE

3 to 4 lbs. sugar
3 lemons
½ lb. raisins
1 slice toast
½ oz. yeast (2 packages)

1. Peel the quinces and put the fruit through a meat grinder. If you have no grinder, chop them into small pieces.
2. Put them into an enamel pot with the water and bring it to a boil, continuing the boiling for forty-five minutes.
3. Strain into a crock, add the sugar, and dissolve it well.
4. Wash the lemons, pare them thinly, and add rinds and juice to the crock.
5. Add the raisins, cut up. Prepare a slice of toast. Dissolve the yeast in ½ cup warm water. Pour the yeast on both sides of the toast and float it on the liquid. Also add the leftover yeast.
6. Cover and let stand in a warmish place (65°-70°) for fourteen to twenty-one days, until all fermentation has ceased.
7. Strain the wine into glass jars. Let it rest for two weeks or more, and then siphon off the part that is brightly clear. Fine the rest with eggshells or isinglass. When the wine is all clear, siphon it into bottles, and do not drink it for at least a year.

This is a slow-maturing wine and needs time. But if you have the patience, you will get a good, semi-dry wine.

This recipe is for one gallon of wine. To make two or more, multiply all the ingredients.

 Raisin Wine]

NO BOOK OF FOLK WINES is complete without recipes for raisin wine. It was the wine I first drank, besides kümmel wine or liquor or brandy (I don't know just what it could be called), made in our home in Austria by my mother. Raisin wine was perhaps the only wine made and drunk by Jews in whatever lands they lived, by choice or by forced migration. The wine made at home from raisins was less expensive than purchased wines, apparently more pleasant to drink, and was used in some of their ritualistic ceremonies, such as Passover.

Raisin wine was one of the most popular in England and was also widely drunk in Italy and France by the natives. I saw it made both in Europe and here in America in very unscientific traditional ways, with the most excellent results. Each household made its raisin wine according to the recipe of the particular family.

The procedure, as with all folk wines, is very simple, and you can get either a dry, sherry-like wine or one as sweet as a muscatel. Here are several recipes.

I

You need: *3 to 6 lbs. raisins*
3 lemons
2 gals. water
6 to 8 pounds sugar
½ oz. yeast (2 packages)

1. Wash the raisins in cold water to remove the chemicals on them, cut them fine, and put them into a crock. (In Austria, they were put in uncut.)
2. Add the thinly pared rinds of well-washed lemons and also their juice. (No lemons were used in my early days.)
3. Heat the water, in which you have dissolved the sugar, and pour it over the raisins. Cool to lukewarm.
4. Dissolve the yeast in ½ cup lukewarm water and pour it on the liquid. Cover and let ferment for ten to fifteen days, when the fermentation should stop. Strain the wine into glass jars and let it rest for two to three weeks.
5. Siphon off any wine that is clear; fine the rest with eggshells or isinglass. Let it rest again for a week before bottling. Later, the bottles will probably have to be decanted again because there will be some sediment at the bottom. However, you can wait with this decanting until the wine is to be drunk. This can be two months after it is finished, but the older it is, the better it will taste.

II

Here is another simple way to make your raisin wine.

You need: 3 gals. water
5 to 6 lbs. raisins
5 to 8 lbs. sugar
½ oz. yeast (2 packages)

1. Put the water into an enamel pot and add the washed, chopped raisins.
2. Bring the water slowly to a boil and let it simmer for almost two hours.
3. Strain it into a crock through a sieve or a fine colander or even gauze. Squeeze every drop of juice out of the raisins.

4. Now put in the sugar, depending on how sweet you want the wine to be, and dissolve it.
5. Follow steps 4 and 5 in the preceding recipe.

III

In the northern part of France I tasted a raisin wine made with pure apple cider instead of water. It needed no yeast, but a sliced lemon was added for each gallon of wine made. After a week of fermentation, it made a marvelously warming drink for a cold day.

IV

Raisin wine can also be made without sugar. In this recipe you need 8 pounds of raisins for each gallon of water.

The raisins should be allowed to rest in the water for a month, stirring them once every day.

Drain off the must. It will keep on fermenting for as long as two or three months. (If I remember correctly, this is how it was made in Austria when I was a child.)

Then add 5 pounds of honey and a quart of brandy for every 2 gallons. You will have a wine to make you walk on air and savor your food with great gusto.

V

There are a thousand and one ways of combining raisins with homemade wines. As a matter of fact, almost all homemade wines have raisins in them, partly because of their flavor and partly because raisins are an aid to fermentation. They can be combined with any kind of berries, fruits, cereals, or flowers. My son made a raisin and prune wine from a recipe I gave him, which tasted so much like a sherry that no ordinary wine drinker could tell the difference.

This recipe was given to me by a Vermont lady at Cooperstown.

You need: 1 lb. raisins (*I used 3 lbs.*)
              1 lb. prunes (*I used 3 lbs.*)
              1 qt. raspberries (*I did not use any*)
              3 lbs. sugar to each gallon of water
              4 qts. boiling water (*I used 8 qts.*)
              ½ oz. yeast (*2 packages*)

1. Put everything into an enamel pot except the yeast, and heat well.
2. Pour into a crock, and when lukewarm, add the yeast, dissolved in ½ cup warm water.
3. Let the must ferment for three weeks, then strain and clear in the usual manner.

You can combine raisins with other fruits or flowers you like, using about 50 percent raisins (by volume).

VI

I cannot help adding a recipe that was common around 1700. You can use it today, and it will give you a good wine.

    "The most approv'd way to make
       Raison Elder-Wine.

"Take six gallons of Water, and boil it half an Hour; and when 'tis boil'd, add to every gallon of Water five pounds of Malaga-Raisins shred small; pour the Water boiling-hot upon them, and let it stand nine Days, stirring it twice a Day: Boil your Berries as you do Currants for Jelly, and strain it as fine; then add to every gallon of Liquor a pint of Elder-berry-Juice: When you have stirr'd all well together, spread a Toast on both sides with Yeast, let it work a Day and a Night, then put it into a Vessel, which be sure to fill as it works over; stop it close when it has done working, 'till you are sure 'tis fine, then bottle it."

 *Rhubarb Wine*]

THIS FRUIT OF THE EARTH with its red and green stalks and broad green leaves has sometimes been called the "wine plant" because of the unusually excellent wine that can be made from it—and I can attest to the truth of the statement. It has also been used in medicine, and during the eighteenth century it was very popular in France for many ailments.

During the Black Era of the United States, Prohibition, my brother made rhubarb wine on a large scale for his personal use and pleasure, as well as for the pleasure and joy of his friends. He had a big wine press, 20- to 50-gallon barrels, and so there was plenty for all. And the fine, golden-yellow, dryish, near-Graves truly brightened the meals in those black-bigoted days. The rhubarb wines I have made lately are of equally good quality. I have been told that British folks also realize what a fine drink can be made from this plant, and it was, and is, a very popular drink because of its vinous delight and medicinal qualities.

There are many ways of making rhubarb wine, but fundamentally they are all alike. The fruit combines well with other flavors: tomatoes, rose petals, any kind of berries or the juice of berries, raisins, plums, oranges, figs, dandelions, marigolds, and the like. In each case, the fruit or flower is added to the rhubarb, and the rest of the procedure is the same as for plain rhubarb wine.

The Recipes

I will set down here the fundamental recipes. From these you will be able to make any variation you like. One of the recipes tells how to make the wine with yeast, and one without yeast. Since I favor making wines with yeast, I will begin with it.

I

You need:
- 10 lbs. rhubarb
- 2 gals. water
- 6 to 8 lbs. sugar
- 1 lb. raisins
- 3 lemons
- 2 oranges
- ½ oz. yeast (2 packages)

1. Wash the rhubarb stalks thoroughly but do not scrape them. (If home-grown, you can just wipe the stalks.) Cut into pieces 1 to 2 inches long and put them into a crock.
2. Boil the water and pour it over the rhubarb.
3. Squeeze or pound the rhubarb until it is mushy.
4. Let it stand for a week, stirring and squeezing it once each day.
5. On the eighth day, strain the juice through a thick cloth into an enamel vessel, squeezing every drop of liquid out of the rhubarb stalks. This is not easy, and if you have a small hand press, it will be of great help.
6. Now add the sugar, dissolve it, and then heat the liquid.
7. Pour it back into the crock and add the raisins, cut up the best you can; the lemons and oranges, washed and sliced; and the yeast, dissolved in ½ cup warm water. In a day it will begin bubbling and dancing to please the heart of any good home vigneron. Fermentation takes fourteen days, more or less.

8. Strain the wine into glass jars, let it stand for two weeks, and if not perfectly clear, fine it.

The wine should be a bright golden-yellow topaz, clear and brilliant. After a year, longer if possible, you should have a magnificent drink. Never drink the wine before it is one year old!

II

My brother made his rhubarb wine without yeast, as home wine makers have done for hundreds of years. It is very simple.

You need: *10 lbs. rhubarb*
2 gals. water
6 to 8 lbs. sugar
3 lemons
3 oranges

1. Follow step 1 in the preceding recipe, crushing the fruit with a wooden masher as thoroughly as possible.
2. Boil the water, pour it over the rhubarb, and continue macerating.
3. Cover well and set in a warm place, mashing and stirring the must once every day with a heavy wooden utensil. On the first day add the sugar, dissolving it thoroughly. Fermentation should start within two or three days. If it does not, add ½ ounce yeast dissolved in ½ cup warm water.
4. Let it stand, fermenting, for fourteen days. Then strain it into an enamel pot, squeezing every drop of juice from the rhubarb stalks. Pour the juice back into the crock and add the thin rinds and juice of the lemons and oranges. Let the must stand for another fourteen days.
5. Strain the wine into glass jars and let it rest for at least two weeks. By the end of that time, the wine, or most of it, should be shining clear. Siphon off the clear part into bottles and fine the cloudy part. Do not drink this wine for at least a year.

III

A nice sherry-like wine can be made from rhubarb by the addition of dried prunes, provided you let the wine rest for a year or two.

You need: *4 lbs. rhubarb*
            2 gals. water
            1 lb. raisins
            2 lbs. prunes
            2 lemons
            6 lbs. sugar
            ½ oz. yeast (2 packages)

1. Wash the rhubarb, if bought; just wipe it, if it is homegrown. Then cut it into 1- or 2-inch pieces and put them into an enamel pot containing the water.
2. Add to it the raisins, cut fine, and the prunes, but be sure to rinse them first in cold water, unless you buy those to which no sulphur has been added.
3. Heat the liquid to boiling and let it boil for a full half hour, mashing it a few times with a wooden spoon.
4. Strain the liquid into a crock, squeezing the juice completely from the fruit. Add handful of prune pits, some of them cracked open, and their kernels.
5. Wash the lemons, pare them very thinly, and add the rinds and juice. Now put in 6 pounds sugar (8 pounds if you want a sweeter wine) and dissolve it.
6. Dissolve the yeast in ½ cup warm water and pour that in. Cover and let stand in a warm place (65°-70°) to ferment.
7. When fermentation has ended, strain the wine into glass jars and let it stand for two weeks. Siphon off the clear wine and fine the rest. This makes an excellent sherry-like wine.

You can also make a champagne-like wine by adding a lump of sugar or some raisins to each bottle, using strong wine bottles and corking with champagne corks with wire cages over them.

FLOWER WINES

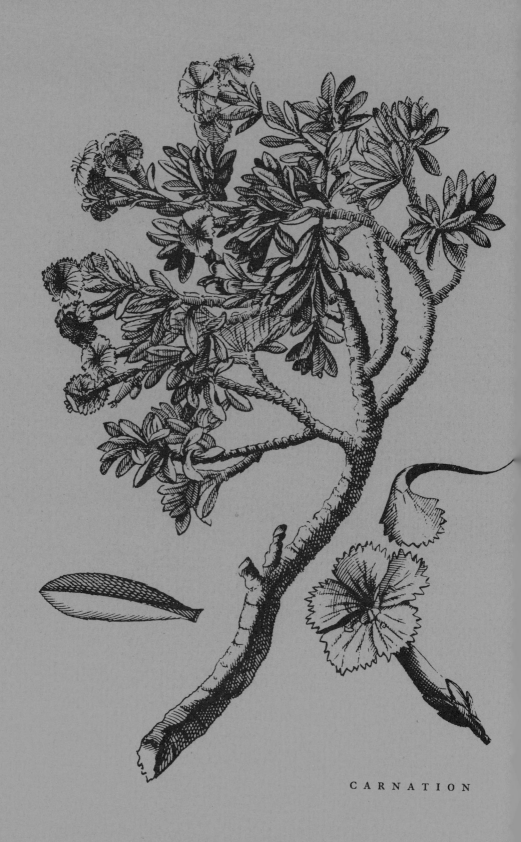

CARNATION

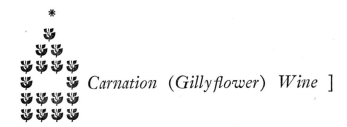 *Carnation (Gillyflower) Wine*]

I REMEMBER READING in a book of medieval lore that a carnation was one of the first flowers to come on the earth after the birth of Christ, and that the pinks were born from the tears shed by the Virgin Mary on her way to Calvary. With such an origin, and the lovely, Oriental scent it gives, how can you help but desire to drink a wine made from this flower? From the earliest days, it was associated with wine. Wasn't its popular name "sops-in-wine"? Thus Spenser speaks of the gillyflower, and thus, too, the greatest of all poets, Shakespeare.

Theophrastus, writing in the fourth century B.C., named the flower dianthe, meaning "the flower of the gods." And the name dianthe has the song of the flower in it.

The Greeks and Romans used carnations for garlands, calling them the flowers of Jove. There is unquestionably some more remote religious meaning in this.

Miss Buckner Hollingsworth, in her excellent and erudite book, *Flower Chronicles*, points out that there are hundreds of years of silence about the beautiful, delightfully scented flower. There may have been silence, but there certainly was usage. Popularity lies in years of growth and habit, and for the carnation to have become one of the favorite flowers of painters during the Renaissance was surely the result of years of familiarity. So we learn that the carnation was used as a medicine "to cheer the heart." It became the favorite flower of poets during Chaucerian days.

In olden days, wines, beers, ales, mead ("metheglin"—if you prefer the old name), and hippocras were flavored with carnations and pinks.

The flower was widely used in cooking. The good cook of the Queen Henrietta Maria made a most excellent wine from pinks.

They were used extensively in foods, sauces, as a conserve, and in many "divine cordials." For hundreds of years the carnation—pink—gillyflower—was one of the most valued flowers in England.

I have had a special personal fondness for this Eastern-scented, flowing white-and-crimson frondescence. I wore a carnation (or a rosebud) in my buttonhole every day for many, many years, and the beauty and scent of the flower enriched my day's work.

With such a history and the word's meaning—in the language of flowers, "pride and beauty"—it also must hold a high place in viniculture. It makes a wine that will go violing on your tongue and down your throat to bring you a song of warmth and well-being.

The Recipe

You need: *2 qts. full-grown carnations (heads only)*
2 gals. water

6 to 8 lbs. sugar
2 lemons
2 oranges
1 lb. raisins
1 slice toast
½ oz. yeast (2 packages)

1. Take full-grown, well-scented carnations, bought or home-grown, and put them into a crock.
2. Boil the water in an enamel pot and pour it over the flowers. Let this stand for four days, stirring it daily.
3. On the fifth day, strain, squeezing the liquid out of the flowers through cheesecloth, into an enamel vessel.
4. Dissolve the sugar completely in the liquid.
5. Pare the lemons and oranges thinly. Squeeze the fruit and add the rinds and juice to the liquid.
6. Cut the raisins into small pieces and put them into the liquid. Now heat it to lukewarm and pour it back into the crock.
7. Prepare a slice of toast, dissolve the yeast in ½ cup lukewarm water, and put the yeast-covered toast into the crock. Cover and set it in a warm place (65°-70°) to ferment. This will take from ten days to three weeks.
8. Strain, fine, bottle, and cork.

The wine should not be drunk for ten months; a year or two is even better.

[*Clover-flower Wine*]

THERE IS A GREAT BODY of legendary tales and beliefs centering around summer-smelling clover, and this alone should make it worthwhile to make a wine from it, without mentioning the virtues of the beverage.

Today clover is mainly associated with cattle feeding, save in Ireland. There, the four-leaved clover grows where fairies have trod and held their midnight dances.

Of course, the clover has no end of traditions in other lands. There is the holy clover of Bethlehem, for such was the bed of sweet-scented grass in the manger where the infant Jesus lay.

In Latin, clover is *clava* (club), in reference to the three-headed club (three-leaved) that giant Hercules carried along on his superfantastic exploits. And clover leaves, like so many other leaves, were used for divination.

In the Middle Ages, jousting knights wore clover on their sleeves for good luck and to ward off evil; and they wore it in battle so they would come out unharmed. Peasants carried clover on their clothing and bodies for the same reasons.

Clover also invaded the realm of medicine. It was a sovereign cure for lunacy, among other ailments. Women wore the little green leaves to promote fertility. For long-running years, unto today, the four-leaf clover is synonymous with good luck.

The three-leaved little plant, the shamrock clover, has a great legendary life in the lore of Ireland, where it is the emblem of the land. To the Irish and many others, the three leaves and stem form the cross that brings good luck and keeps away misfortune.

Put a four-leaf clover under a pillow, and you will sleep pleasantly and dream of the one you love or the one who loves you. Some good Irish lads even believe that if you carry the four-leaf clover hidden somewhere on your body, the draft sergeant will never nab you. Perhaps more charms are associated with this than with any other plant.

I know some folks who have a miraculous faculty for finding the lucky leaves. There is our Rosita, from Colombia, half Indian and half Spanish: all she needs do when walking in the green grass is to bend down, and up she will rise with the so much sought-after four-leaf clover. But I, ever on the lookout for it, have found only one or two in all my life. Which goes to prove: "There's nought so queer as folks."

The Recipe

Of course, you need a place in the country where you can pick the clover. But if you do not have your own, you can easily find a farm where it grows and where you will be permitted to pick the flowers, if you tell the farmer for what purpose you want them.

You need: *6 to 8 qts. clover flowers, fairly tightly packed down*
           2¼ gals. water
           5 to 6 lbs. sugar
           4 lemons
           4 oranges
           1 oz. fresh ginger root
           ¾ oz. yeast (3 packages)

1. Gather white or purplish clover flowers—the color does not matter—and spread them out on a sheet in the shade to dry.
2. Put the water in an enamel pot.

3. Dissolve the sugar thoroughly in the water.
4. Wash and peel the lemons and oranges thinly. Put the rinds into the liquid, then squeeze the juice of the fruit into it.
5. Put in the flowers and the ginger.
6. Bring slowly to a boil and let it simmer for a good half hour.
7. Now strain it into a crock and let it cool somewhat.
8. Dissolve the yeast in ½ cup warm water. When the liquid is lukewarm, spread the yeast on top.
9. Cover and let the must ferment in a warm place (65°–70°). It will take ten days or more. When the wine is still, strain, clear, fine, decant, and then bottle.

This wine needs at least one year to mature.

 Cowslip Wine]

THE AMAZONIAN JUNGLE does not compare with the confusion in the expressive names of plants. There are Latin names and local names and folk names and country names and other names. The cowslip stands high in this Babel-ish medley. Suffice it for us common mortals, not deeply initiated into the mysteries of these names, that we have the cowslip in America. It is a perennial herb belonging to the primrose family and is quite common in the Middle West, where it is often called the "pride of Ohio" or "shooting star."

The Greeks, who in ancient days created beauty of any and every thing—as the Japanese do today—drew from the spring-scented fragrance of this flower the garlanded, lovely tale about Paralisos, the son of Flora and Priapus, who died of a broken heart when the one he loved was turned into the delicate, fragrant primrose.

Our cousins, the British, revel in the cowslip, often called primrose, which ranks very high in their country. I might almost call it the national flower of England. It is considered a fairy flower, with endless virtues that have endeared it to folks for centuries. In some parts it is known as "fairy cup" as well as "fairy flowers," for in the cowslips fairies find shelter from storms and rains. The cowslip was also called the "wine-producing flower." According to the great poet, Pope:

". . . If you need a rest,
Lettuce and cowslip wine probatum est."

It has figured endless times in the poetry of England.

". . . Where the bee sucks, there suck I;
In a cowslip's bell I lie."

And again the great bard of Avon:

". . . And I serve the fairy queen,
To dew her orbs upon the green,
The cowslips tall her pensioners be;
In their gold coats spots you see;
Those be rubies, fairy favors;
In those freckles live their savors.
I must go seek some dew-drops here,
And hang a pearl in every cowslip's ear."

In flower language, also, it is the fairy flower, with the divine power of preserving continuous youth or restoring it when lost.

In folk medicine, too, it holds its place. *A Book of Fruits and*

Flowers, 1653, states: "Cowslip conserve [for which there is an excellent recipe] doth marvelously strengthen the Braine, preserveth against Madnesse, against the decay of memory, stoppeth Head-ache, and most infirmities thereof." From the days of the Greeks, and probably before, it has been used as a gentle narcotic. Cowslip tea taken before bedtime will help the sleepless.

In short, what the tulip is to Holland, the cowslip is to England. Good Queen Victoria and her ladies, when at the palace of Osborne, gathered lovely, scented primroses and sent them to Prime Minister Disraeli in London. I wonder what lovely, scented thoughts wandered around the little Queen's mind while she gathered the flowers.

Look at the blooms, and you will see that the flower cluster resembles a bunch of keys hanging on a stalk. So one of its names was "Saint Peter's-wort," because Saint Peter was always pictured carrying the keys to Heaven.

There is a charming tale connecting Saint Peter with these flowers. It comes from Germany, where they are called "keys of Heaven." One day good Saint Peter heard that a black, sinning soul was trying to sneak into Heaven through the back door. He became so angry, excited, and flustered that he let the keys fall out of his hands. They dropped down to earth, and where they fell, primroses grew. Thus the name. Since they were heavenly keys, the flowers were gifted with the power to open locks of treasures and caves.

In the early days, there was no greater favorite than the cowslip for use in beauty potions. Cowslip beauty cream reigned supreme, for ". . . it taketh away the spots [freckles] and wrinkles of the skin and adds beauty exceedingly." But Mr. William Turner, the bluenose of those days, writing in *A New Herball*, held forth against the innocent flower. Listen to his boiling fury: "Some women sprinkle ye floures of Cowslip wt whyte wyne and after still it and washe their faces in that water to drive wrinkles away and to make them fayre in the eyes of the worlde rather than in the eyes of God whom they are not afraid to offend with the sluttishness, filthiness and foulness of the soule."

The flower was also a great favorite in cooking, and there are endless recipes of how it was used in conserves, pickles, tarts, puddings, syrups, and the like. Cowslips also played a part in children's games. The flowers were rolled together into a ball and then used in a game with the tangy name of "tisty-tosty" (which, incidentally, was quite an old game). And in the olden days, it was a flower much used for making folk wines. Perhaps the fact that the flowers bore the names "dunkards," "crazy Bet," "publicans and sinners" is a reflection of this usage. And now to the recipes.

The Recipes

I

In bygone days, a "smoothe meade" was made of cowslips, lemon juice, honey, and a handful of brier. It was a popular and well-known drink in England.

There are many, many recipes for wine making, using the American cowslip in the same fashion as the English cowslip.

I'll begin with an old, old recipe that was set down by the famous John Evelyn in his *Acetaria: A Discourse of Sallets* (1699). It is so simple that you can make it today.

"To every Gallon of Water put two Pounds of Sugar; boil it an Hour, and set it to cool: Then spread a good brown Toast on both Sides with Yeast: But before you make use of it, beat some Syrop of Citron with it, an Ounce and a half of Syrup to each Gallon of Liquor: Then put in the Toast whilst hot, to assist its Fermentation, which will cease in two Days; during which time cast in the Cowslip-Flowers (a little bruised, but not much stamp'd) to the Quantity of half a Bushel to two Gallons (or rather three Pecks) four

Lemons slic'd, with the Rinds and all. Lastly, one Pottle* of White or Rhenish Wine; and then after two Days, tun it up in a sweet Cask. Some leave out all the Syrup."

II

Cowslip wine is a mild soporific, and I highly recommend a small glass before bedtime. It is perfectly harmless, and a most pleasant habit.

You need: *4 qts. cowslips (heads only)*
3 lemons
3 oranges
2 gals. water
8 lbs. sugar
¾ oz. yeast (3 packages)

1. Gather the cowslips. If the flowers are moist, dry them. Put them into a crock.
2. Wash and peel thinly the lemons and oranges, and put the rinds in with the flowers. Then squeeze the juice of the fruits and put that in.
3. Boil the water and sugar in an enamel pot for about five minutes, removing any scum that may rise. Then pour the syrup into the crock. Let it stand until it is lukewarm, stirring it vigorously for a time with a wooden spoon.
4. Dissolve the yeast in ½ cup warm water and pour it into the crock.
5. Cover, and set it in a warm place for five or six days, stirring it once every day. At the end of that time, strain the must into glass jars or gallons and let it continue fermenting.
6. When the fermentation is finished, strain, decant, fine if necessary, and then bottle.

* A pottle = ½ gallon.

This will make a delightful dry wine with a delicate bouquet, and the longer it is kept, the better it will become.

III

Here is a somewhat different recipe, which infuses a slightly different, piquant flavor into the wine.

You need: *4 lemons*
4 oranges
1 handful sweet brier or 1 handful pineapple-sage leaves or 1 oz. fresh ginger root
4 qts. cowslips (flowers only)
2¼ gals. water
4 lbs. honey
¾ oz. yeast (3 packages)

1. Put into a crock the thin rinds and the juice of the lemons and oranges.
2. Add a few twigs of sweet brier, or a handful of pineapple-sage leaves, or ginger root.
3. Gather the cowslips; be sure they are dry and put them into the crock.
4. Put the water in an enamel vessel with the honey. Bring it slowly to a boil and let it simmer for thirty minutes, then pour the boiling syrup into the crock.
5. When lukewarm, add the yeast (as in recipe No. 2) and let the must ferment. At the end of the visible fermentation, strain off into glass jars and then proceed to clear.

If given a year or more to rest, I know of no more pleasant nightcap to bring pleasant dreams.

 Daisy Wine]

THAT, of alle the floures in the mede,
Than love I most these floures
  whyte and rede . . .
Allas! that I ne had English,
  ryme or prose,
Suffisant this flour to preyse aright!"

Thus great Chaucer sang of the daisy.

The flower of fortunetelling and innocence! In my boyhood days I plucked the slender white petals to learn if the one I loved, loved me. I questioned also—and I hope you did too—the same long white petals to learn if I would be a "rich man, poor man, beggarman, thief," or "doctor, lawyer, Indian chief." The oracle came just as true as if given by a learned, modern, heavy-charging psychiatrist.

The ancients used daisies for the same purpose—for oracles—and also for making wreaths. The flower was then an emblem of fidelity, and to this day it is still, in a measure, a symbol of innocence, so glowingly illustrated at one of our great women's colleges, Vassar. There, from the ancient echoes, it is wreathed and borne by lovely, flushed faces and lovely, lissome bodies as an emblem of innocence. This is strangely incongruous, considering the actual coarse entry of the flower into our land, with General Burgoyne—as fodder for the horses. No, not fat Burgoyne, but the innocent daisies.

This flower has been called the "day's eye," for it closes up gently at night and opens its eyes at the roseate dawn of morning. So precious was it considered by ancient peoples that even four thousand years ago it was the model for a design for long, beautiful golden hairpins. These have been found in a Minoan palace.

The white-petaled golden-disk flower was used as an ornamental border in mosaic work by Assyrian and Egyptian craftsmen.

Like all growing things, it was part of the medicine of the people, particularly for the eyes: the daisy's center looked like a yellow eye. So when the newer belief arose that every flower was a written word of God, so ably expounded by the alchemist Paracelsus, in his *Theory of Signatures; God's Hieroglyphickes,* the "eye" of the daisy indicated its purpose in life—to be used in many ways for the eyes. It was Pliny, in his *Natural History,* who first said the flowers are expressions of the gods.

Later, daisies had many more medicinal uses: Pour ale and holy water over the flowers, sing a certain rhyme, and baneful sores will vanish. That was a well-known remedy.

With time, daisies could cure almost anything: warts, insanity, the plague, smallpox. They truly became a medicine for "all kind of hurtful aches in whatsoever way they came." Most important of all, a daisy would foretell if the patient would live or die. "Take the flower of the Daisy and pound it well with Wine, giving it to the patient to drink; if he vomits it, he will die of the disease, if not, he will live and this has been proven."

Since it was an emblem of innocence, many ladies used it in their coats of arms.

I could write pages about the names given to the flower. The French name, marguerite, means pearl—this from the whiteness of the petals. There has been a galaxy of queens and princesses so named. Saints bore the name: Margaret of Antioch, Saint Margaret of the Dragon, who was driven from home by her father because she would not renounce her new belief in Christianity. She ever kept her face turned to Heaven in prayer and meditation. "Maid Marguerite, meek and mild of Antioch" has been a great helper of women about to become mothers.

Its botanical name, Bellis, conceals a charming story. It is derived from the Belides, the dryads, wood spirits. While one of them was dancing, she was seen by Vertumnus, the God of Spring. He ran with winged feet to clasp her in his arms. White with fright, she prayed to the gods and sank to the earth—and they formed her into a daisy.

The popular little flower naturally had many local names: "little Easter flower," "trembling star," "measure of love," "a thousand charms," and many others.

No wonder the sturdy daisy has been used endlessly in poetry, folklore, proverbs, and customs. The greatest English poets, from Chaucer down, wrote of it. In England, spring had arrived only when your foot could cover twelve daisies. Sleeping with a daisy root under the pillow and your shoes hanging out of the window would surely bring your lover. If you dream of daisies in the summertime, it is good luck, but be careful! If you see them in your dreams in the wintertime, it means ill luck.

In florigraphy, daisies meant purity and youthful bloom. A daisy sent to one in our land meant "April, and I will think of you."

Thus the daisy has a large place in lore, in life, and in literature. And it should have a place in viniculture.

The Recipe

Many old recipes do not include the use of yeast in the making of daisy wine, implying that the flowers have enough yeast in themselves for fermentation. I tried it; it was a rank failure. Then I made it again, this time with yeast, and it was successful. Wherefore, I give you a recipe with yeast.

DAISY

You need: *8 qts. daisies (heads only, packed down fairly hard)*
2 gals. water
6 lbs. sugar
1 lb. raisins
4 lemons
4 oranges
½ oz. yeast (2 packages)

1. Put the daisy heads into a crock.
2. Boil the water in an enamel pot and pour it over the daisies.
3. Let stand for twenty-four hours, mashing the flowers now and then with a wooden spoon.
4. Drain the liquid into the enamel pot, leaving the daisies in the crock. Dissolve the sugar thoroughly in the liquid.
5. Add the raisins, cut finely; the washed, thin rinds and the juice of the lemons and oranges; and the juice of the daisies left in the crock, squeezed through a cloth.
6. Heat the mixture slowly; let it simmer twenty to thirty minutes.
7. Pour it back into the crock and when lukewarm, add the yeast, dissolved in ½ cup warm water.
8. Cover and stand in a warm place (65°-70°) to ferment. This will take from fourteen to twenty-one days. At the end of that time, strain into glass jars or gallons. Clear and fine, if necessary.

If you let it mature for one or two years, you will have a nice, dry, Graves-like wine.

This is one of the three wines of the many I have made that has a tendency to fizz and become champagne-like. The other two are wines made from elder-flowers and rose petals. It is therefore advisable to use a heavy Burgundy or champagne bottle for these. It would be advisable—and this is what I have done—not to seal-cork daisy wine until it becomes completely still, letting the bottles stand upright during this period.

[Dandelion Wine]

OF THE ENDLESS RECIPES I received from folks and friends from one end of the land to the other, the most numerous were those for the oracular dandelion. It unquestionably can be ranked as the most popular folk wine made in our country.

I call it the oracular flower, for to it are attached more divinatory attributes than to any other.

It is also the only flower, as far as I know, for which a fleet was named.

When I was collecting folktales of New England,* I came to Gloucester, and there a fisherman spoke of the "dandelion fleet" around the 1890's. This name was given to the fishing vessels that, after a winter's inactivity, set out for their catch when the dandelions began to bloom.

The dandelion also has the distinction of being the one flower that bears the same name in all lands. "Tooth of the lion": "*dent de lion.*" The exact origin of the name is dubious. Perhaps it is so called because its leaves resemble the jagged shape of a lion's tooth; perhaps because they resemble the golden teeth of the heraldic lion; or perhaps because the yellow center is a tiny golden sun, the lord of warmth and light of the world, even as the lion is the lord of the forests and fields.

In some parts of England I heard that now and then—but rarely

* *New England Bean Pot.* Vanguard Press, New York.

—the flowers were called "swines-snout." The origin of all these names is deeply buried in the vast repository of time.

And speaking of time, dandelions were used and eaten in Biblical days even as they are today. Many, many tales are associated with them. My favorite is the one told by the Algonquin Indians of New York. The Indians wove a truly golden tale around the dandelion and the South Wind, Shawondasse. To me, some of the North American Indian words have a music the like of which is not found in other languages. Where do you find a more melodious-sounding word than Monongahela? It needs three English words to produce a similar effect, such as a "blue morning glory."

Shawondasse, the South Wind, lazed among the trees and in the plain, and there he saw a beautiful golden-haired maiden. No sooner did he see her than he loved her. But he was a lazy fellow, too lazy to go to speak to her.

He saw her every day for many, many days, but he was always too lazy to tell her of his love; he would just gaze at her golden face and hair. Then one morning he saw a strange sight. There was the maiden, but her golden hair had become all gray and white.

"My brother, North Wind, touched her golden hair with his ice-cold hands and turned it white," he said.

He felt sad and heaved a slow, deep sigh. So strong was the sigh, it reached the white hair—which, wonder of wonders, flew away with the wind, scattering in all directions.

And this happens every year—for South Wind does not remember and is always lazy.

I must add here the quaint Greek tale in which the dandelion has a unique place.

Pirithoüs thought it was time for him to marry, and he decided he would have none other than one of Jupiter's daughters, Persephone, the Queen of the Netherworld. That she was married to Pluto did not concern him. So he asked Theseus to go with him to the Netherworld to carry her off.

But Pluto learned of the plot and lay in wait for them. He seized them both and set them on an enchanted rock on which they were held by a magic spell.

Persephone knew of the two adventurers, even as the whole Netherworld did, and when Pluto was away, the Queen, prompted by woman's curiosity, went to see the two prisoners. She looked at Pirithoüs and at Theseus, and Theseus found favor in her eyes. Taking pity on them, she smuggled to them each day a fine dish of . . . dandelions and onions! This she continued to do for a long time—until Hercules descended to Hades for the three-headed dog and freed Theseus.

The dandelion flower can truly be called the flower of fortune-telling. The young in years and spirit have solved more mysteries and problems of life with the humble little golden sun than with any other flower.

When the flower has changed to seed, blow on the "blowball," as it is then called, and you will know the time of day by the number of times it takes you to blow off the seeds.

The number of times you blow on it will also indicate how many times you will be married.

If you blow on a blowball three times, and some of the seeds have not blown away, the one you love is thinking of you. The remaining seeds will also show how many times you will be married and how many children you will have.

The same three breaths, with a single seed remaining, means that your wish will positively be fulfilled. Blow on the blowball, and the number of seeds that fall on your clothes tells how many sins you have committed.

Whisper your love message to the blowball, then take a deep breath and blow on it, and the white-tufted messengers will fly where your lover lives and convey your tender words.

But if you dream of the dandelion, misfortune is on its way to you. And when the blowball silk flies of its own accord, rain is on the way. Some call it the witch's plant. Remember that yellow is a symbol of gold in folklore land.

The dandelion is also a plaything for children. The stems can be curled for necklaces and bracelets, and I have seen them worn as earrings.

The use of the dandelion as a clock is common among folks in many lands. In Switzerland, it is the "shepherd's clock"; it opens at 5 A.M. and closes at 8 P.M.

Like nearly all flowers, the dandelion has been used in medicine and for food. The renowned Mr. Nicholas Culpeper of England, speaking of the dandelion in his famous book on herbs, has the following to say: "[It is] very effectual for the . . . jaundice, the hypochondrical passion . . . and whoso is drawing towards the consumption, or to an evil disposition of the body. [It is] effectual to drink in pestilential fevers and the sores." It has also been used for many other ailments.

As for food, it was one of the bitter herbs eaten by the Israelites in Biblical days, and has been used in salads ever since. I have picked tender dandelion leaves in the warming spring days for my salad, and I hope you do the same.

It is a lovely Umbrian, pastoral sight to see Italian men or women along our magnificent sunny highways, picking young dandelion leaves for their greens. When cooked, the leaves have a pleasant though bitter taste.

From about 1700 to 1890, the roots of the plants were often dried and used as a substitute for coffee. The roots were also cooked with whey to make a sleep-inducing drink. And this somehow reminds me that a dandelion is a slang term of endearment for your love.

Now, after medicine, food, drink, and loves, let us turn to the recipes for the wine.

The Recipes

As I said at the beginning, I received more recipes for dandelion wine than for any other. Elderberry wine was second in popularity.

I have over a hundred recipes for dandelion wine and am hard put which to set down. But whichever I tell you, whether the one from Ann Grimes, the fine folksong collector and singer from Ohio; or from Bertha Nathan, the Maryland storyteller; or from Horace Hillery, the county historian of my own bailiwick; or from the kitchen of the Fleischmann Yeast Company, or from friends from the South or from Vermont or out West, they are fundamentally alike. I will give only those I have used. You can vary them or combine them according to your taste and pleasure.

There are two good reasons for the popularity of this wine. The first is that dandelions grow everywhere, begging to be picked and turned into wine. The second and more important reason is that the wine of this flower has a uniquely pleasing taste and bouquet. It is a pleasurable drink, apart from any medicinal or stimulating virtues.

So, as a final word, if you have a lawn before your house and the little golden dandelions come up in May looking at you smilingly and confidently, don't swear at them and cry that they ruin your lawn. On the contrary, they beautify it with their golden starry decoration and they tell you quietly: Here we are, ready to be transmuted into a delicious wine that will bring you pleasure and elation and a warming friendship for humanity. All this is very true, for I have been making this wine for the last fifteen years and know whereof I speak.

I

Let me start with a Connecticut recipe, circa 1677, brought by the incoming emigrant from England. I have modified it slightly for today's usage.

You need: 4 to 6 qts. dandelion flower heads
2 gals. water
4 to 6 lbs. sugar
6 lemons
6 oranges
1 tbsp. ground ginger (or ½ oz. fresh ginger root)
3 cups raisins (1 ½ lbs.)
½ oz. yeast (2 packages)

1. Gather the dandelions and put them into the water in an unchipped enamel pot.
2. Add the sugar; bring to a boil and let it boil for a full thirty minutes.
3. Strain; return the liquid to the enamel vessel.
4. Wash and thinly peel the lemons and oranges and put the peelings into the water, also the ginger. Let it simmer for another half hour.
5. Pour the liquid into a crock, and now add the juice of the lemons and oranges.
6. Add the cut-up raisins.
7. Dissolve the yeast in ½ cup warm water and pour it over the liquid when it has cooled to lukewarm.
8. Cover and set it in a warmish place (65°-70°). Fermentation will take from ten to twenty-one days.
9. When the fermentation has stopped, strain the wine into glass jars and let it rest to see if it clears. If it does not, fine it. Then bottle it.

Do not drink it for at least one year.

II

You need: 8 qts. dandelions (*flowers, no stems*)
2 gals. water

3 oranges
3 lemons
4 to 6 lbs. sugar
2 lbs. raisins
½ oz. yeast (2 packages)

1. Pick ripe dandelion flower heads (the stems are never used), wash them thoroughly by letting them stand under cold running water, then put them into your crock.
2. Boil the water in an unchipped enamel vessel and pour it over the flowers. Cover the crock, put it in a warmish place (65°-70°), and let it stand for three days, stirring it with a wooden spoon a few times each day.
3. Strain the liquid into the enamel pot, squeezing every drop of juice out of the dandelions.
4. Wash and peel thinly the oranges and lemons and put the rinds into the liquid.
5. Dissolve the sugar in the liquid. The more sugar you use, the sweeter your wine will be. I like mine drier, so I put in the lesser amount. Bring the liquid to a boil and let it simmer for fifteen or twenty minutes. You can add a little water to make up for the loss in the boiling. Now return it to the crock.
6. Add the raisins, chopped up a little.
7. Add the juice squeezed from the oranges and lemons.
8. Dissolve ½ ounce yeast in ½ cup warm water and add it to the liquid. Cover the crock and set it in a warm place (65°-70°) to ferment. It will take ten to twenty-one days for the fermentation to stop.
9. Strain the wine into glass jars and let it rest for a week or two to clear. If it does not, fine and decant. When absolutely golden, clear it, bottle, and cork. It should not be drunk for a year or two. The longer it rests, the better it will be.

III

The following recipe was sent to me by Dr. Margaret Bryant, the word wizard. She says this is the way folks make dandelion wine in Kansas.

1. Put 4 quarts dandelion heads in a crock; add four sliced lemons.
2. Add 8 quarts hot—not boiling—water.
3. Set away for three days, then put on the stove (in an enamel pot) and boil for twenty minutes.
4. Strain into the crock, add 6 pounds sugar and ½ ounce yeast.
5. Tie a cloth over the top of the crock, and put it in a dark place. When the fermentation has ended, strain and bottle the wine.

When I made it, I set it in a warm place and fined it after the first clearing.

IV

Here are two variations of the preceding recipe, which will alter the flavor of your wine. These recipes are from old cookbooks and from upstate New Yorkers.

A. This one calls for the addition of two peeled bananas to each gallon of water. This will change the orginal flavor slightly. Just remember that you can vary any wine by adding some other fruit or flavor to please your own individual taste. The bananas are added at step 4 in Recipe II.
B. The second variation calls for the addition of ginger, either in the form of powder or the dried root—about 1 ounce—at step 4 in Recipe II. Again, this changes the taste of the wine; it will have a sharper, spicier flavor. I am very cautious about adding ginger, for it can easily obliterate other flavors.

There is still another variation: the addition of a pint of brandy to each gallon of wine when the fermentation has stopped. I do not favor this.

 *Elder-flower Wine*]

I CANNOT HELP but begin this with a snatch of an old Scottish poem I heard in Edinburgh at a happy caille (a true Scottish party with singing, dancing, and drinking).*

> "Mistress Jean she was making the elderflo'er wine;
>   An' what brings the Laird at sic a like time?
> She put off her apron and on her silk gown,
>   Her mutch wi' reid ribbons, an' gaed awa down."

It is a brave beginning for a wine that has been popular for many years.

You might like to know that the elderbush is the only one that will give you both a red wine and a white wine without any additional coloring. From the elderberries come a red, heavy wine; from the blossoms, an exquisitely delicate wine and also a champagne with from 5 to 14 percent alcohol.

Let me say here, to forestall any shock on your part, that it is a tricky wine. I made, the first time, eleven bottles; six of them popped open after I laid them down, though they had shown no signs of fermentation. I rebottled and recorked them, standing them upright. Again the corks blew. Then I put them in champagne bottles, using proper corks and proper wire baskets. And then I had champagne.

* It appears also in Miss F. M. McNeill's valuable book, *The Scots Cellar*.

I don't know why this happened, and I do not guarantee that it will not happen to you. I will try again and see—and perhaps hear!

The Recipes

I

You need: 1 qt. elder-flowers (*heads only*)
            2 lemons
            2 gals. water
            6 lbs. sugar
            1 lb. raisins
            ½ oz. yeast (2 packages)

1. Be sure the flowers are in full bloom and dry when you pick them; then they have a delicate, delightful perfume. Unripe flowers will give the wine a bitter taste. When you strip the flowers off the stems, there is no harm if you leave on the minute green of the florets. It does not spoil the flavor of the wine, but gives it a delicate chartreuse-green tint, pleasant to see. If you have the time and patience, you can use only the corollas. This will take much time, but will give a golden yellow color to the wine. Put all into a crock.
2. Cut the lemon rinds very thin and put them in with the flowers. Then squeeze the juice and add it.
3. Put the water and sugar in an enamel pot. Bring to a boil, then pour it into the crock over the flowers and lemon rinds.
4. Add the raisins, cut as finely as possible, and let the liquid cool.
5. While it is cooling, dissolve the yeast in ½ cup warm water and add it. Elder-flowers, like grapes, usually have enough yeast to produce their own fermentation, so you might wait a day or so before adding the yeast; it may not be needed.
6. Cover, put in a warm place (65°-70°), and let it ferment. Each

day stir the must gently with a wooden spoon. Within ten to twenty-one days, the visible fermentation should end.

7. Strain, squeezing the flowers, etc., through muslin or a kitchen towel.
8. Decant, fine if necessary, and place in glass jars or glass gallons, lightly closed. Let stand so for a month or two—I told you of my experience with elderberries. When you think the fermentation is really over, bottle in thick bottles, first corked lightly. Finally, cork with champagne corks. You can get adequate ones, and they are simple to insert and wire.

In a year or two you will have a fine wine with a delicate rare flavor, or—a champagne wine with the same bouquet. Whichever you get, it will be a reward for your effort and a pleasure to drink.

You can make a good elder-flower mead by just substituting 5 or 6 pounds of *clear* honey, preferably not blended or boiled, for the raisins and sugar; the rest of the procedure is the same.

II

Here is a somewhat different recipe, given to me by a charming lady from Vermont, Mrs. Helen Winston.

Put 2 quarts elder-flowers into a gallon of water (in an enamel pot) and let stand for nine days, stirring every day with a wooden spoon.

Then squeeze the juice from the flowers and let the liquid stand another day.

Now boil it gently from an hour to an hour and a half.

Put in 3 pounds sugar, 2 ounces ginger root, and 1 pound raisins.

When lukewarm, add yeast (½ ounce) on a piece of toast (but see step 5 in the first recipe).

Let it stand until the fermentation stops. Clear and bottle.

Whichever way you make it, you will have an unusually delicately flavored and scented wine, the like of which you have never drunk before. It will be on the sweet side—say like a Château d'Yquem—with a difference.

III

You can set out definitely to make an elder-flower fizzing wine with a very small alcoholic content, which is a pleasant drink on a hot night. It should be served ice cold. It has a delightful bouquet and is greatly favored by ladies who have not become accustomed to burning their taste buds with hard liquor.

Look for the sweet-scented lacy elder blossoms about the month of June, and try it.

You need: *8 to 12 heads ripe elder-flowers*
           2 lemons
           4 tbsps. white vinegar
           2 gals. water
           1 lb. sugar

1. Strip the flowers of their stems and put the flowers into a crock.
2. Add the washed, thinly peeled lemon rinds and also the lemon juice.
3. Add *white* vinegar.
4. Boil the water. If you have spring water or water from an artesian well, it need not be boiled; just pour it into the crock over the flowers. If you boil the water, let it cool first before pouring it into the crock.
5. Dissolve the sugar in the liquid. Use more sugar if you want a sweet wine.
6. Let the must stand for twenty-four hours, then strain it and put it into champagne bottles. Use wired champagne corks.

This recipe does not call for any yeast, since elder-flowers usually contain enough to produce fermentation.

The wine can be drunk four to six weeks later.

 *Goldenrod Wine*]

FRIGHTENING AS IT MAY SOUND to the hordes of hay-fever sufferers, a very excellent golden sauterne-like wine can be made from the so much feared and despised goldenrod. I don't know whether people susceptible to the sight and pollen of this golden-crowned bushy flower waving in the sunshine would also be susceptible to the wine, but it is well worth a trial, if your doctor permits, for the wine is excellent and delicately flavored. And it is very simple to make.

The Recipe

You need: *6 to 8 lbs. sugar*
2 gals. water
4 to 6 handfuls goldenrod (flowers only)
1 lb. raisins
6 oranges
2 lemons
1 slice toast
½ oz. yeast (2 packages)

1. Dissolve the sugar in the water in an enamel pot and bring to a boil. Let it simmer for a few minutes, clearing any impurities that come to the top.
2. Add the goldenrod, cut-up raisins, and sliced fruit.
3. Simmer for five minutes, then pour into a crock.

4. Dissolve the yeast in ½ cup warm water. Spread it over a piece of toast, then add toast and the remaining yeast to the crock.
5. Cover, stand in a warm place, and let it ferment. This will take from ten to twenty-one days.
6. When fermentation ceases, strain into glass jars and let stand for two weeks. If it does not clear, fine with isinglass or eggshells, decant, and bottle. It will be golden in color, pleasant to look at, and excellent to drink.

Marigold Wine]

I WONDER HOW MANY PEOPLE, looking at the lovely marigold, know what an important place it has had for thousands of years in the flower-strewn lore of the world and in men's tales. In addition, this flower has played a great role in foods, in the healing and grotesque medicines of the world, and in wines. There is a wine recipe dated 1671.

Since the marigold is bright and golden in color, poetic-primitive minds naturally connected it with gold. Poets called it the "sun flower," "pride of the sun," the "gold flower," "husband of the sun." There is a charming Greek tale wedded to this idea.

It tells of Catha, the maiden who was so passionately in love with the sun in the skyland that she stayed up all night long without sleep so as to be thrilled by the first sight of her scarlet love. She stayed awake night after night after night for that first moment of passionate thrill at the sight of her puissant golden lord. And so, never sleeping, she slowly pined away and died. And where she stood in passionate adoration and fatal disintegration, the marigold grew.

The marigold was christened with its musical name by monks and nuns of medieval days because it blooms during festivities of the blue-clad Virgin Mary. The florets of its disk are the shooting rays of glory of the Holy Mother, and so it was consecrated to her.

Many flowers were dedicated to the Holy Mother, and there have been many arguments as to which flower holds precedence—particularly among the flowers bearing her name. Some hold that this one was dedicated to the Virgin because she wore it in her bosom.

Of course, there was a much older Roman name, calendula, because the flowers appeared so often on the calends of the month.

Yet in some parts of England it is called "drunkard," for folks said if you gathered many of them, they would make you drunk.

One of the lovely, older uses for the little flower was to make from it an oil that would enable you to see fairies. The recipe is a long one to set down, but if you want to see fairies on moon-drenched summer nights, write me a letter, and I will send you the list of the gall and the other ingredients needed.

It works a great miracle of love for girls in the Balkan lands and may for you, too. The flower is a symbol of undying love. All you need do is dig up the earth where your love walked, put it into a pot, and plant a marigold in it. Then his love will last unto death.

Of course, the marigold was used widely in love potions, and it possessed many magical virtues besides, particularly in England. Worn as an amulet with a bit of laurel wrapped around a wolf's tooth, it would keep folks from speaking ill of you behind your back.

If a man wore it on his body, he was safe from any kind of peril, be it in the battlefield or in the home.

Something stolen from you? Put a marigold around your neck, and it will tell you who the thief was. And if you are in cankered difficulties with your wife, go with a marigold to the next Mass, and she will become magically sweet and friendly. If you dream of this flower, it portends a happy, rich marriage, wealth, and success.

That is only a beginning of its multiple virtues. Here are a host more, enumerated by William Langham in *The Garden of Health*.

"Drinke the flowers with wine to comfort the stomach, procure appetite, to consume humours of the stomach . . .

"Bruse a Marigold leafe, and put it into the nose to cleanse the head and avoid reume.

"Take the Conserve of the Flowers fasting, to cure the trembling of the heart, and to withstand the plague and euill ayre.

"Wash the mouth therewith for the toothache, it is a plesent remedie.

"To preserve you from death, drink 111 grains of Marigolds.

"Wits lost or mad, take the juyce of Marigold and Wormewood, of each one spoonfull."

There are many more, too long to enumerate.

The marigold is also one of the flower weather prophets. It closes its leaves when storms come roaring on the winds. It turns its leaves toward heaven when the full sun comes along, and closes them when dark night descends.

So Shakespeare sang:

"The marigold, that goes to bed wi' the sun
And with him rises weeping."

Medically, the marigold was used by the Assyrians and on down. Later, it served to cure humors in the head (how much more musical, that name, than "migraine headache"!) and almost everything else—inflammation of the teeth, inflamed eyes, the plague, dewy melancholy, and the heartaches of racking love. These are just a few of the many maladies this little flower would assuage.

The golden marigold was brought to our North America by the first settlers, and soon spread over the land. It was used for brocaded ornaments in gardens; and in foods and medicines, some say,

MARIGOLD

WILD PANSY

even as late as the first World War, when it was applied as a hemostatic.

In Mexico, marigolds are sometimes called "death flowers," for they grew from the earth stained with the blood of the Indians the Spaniards tortured and killed in their lust for gold.

In food, marigolds are served in many ways—as flavoring with other herbs, in meats, salads, and broth. Here is a recipe proposed by John Nott's cook (eighteenth century) to the Duke of Bolton, which you can try in your home this summer if you have an herb garden.

"Mince several sorts of sweet herbs very fine—Spinache, Parsley, Marigold flowers, Succory, Strawberry and Violet leaves. Pound them with oatmeal in mortar. Boil your oatmeal and herbs in broth and serve." It should make a fine soup. Of course, you can add or substitute any edible herbs you like.

Marigolds will give exotic flavor to puddings—I can attest to that—and fine "green Pudding" (sweet meat dumplings) can be made with them, as was already noted in *The Compleat Cook* of 1655. Here is the good recipe:

"Take a penny loafe of stale bread. Grate it, put to halfe a pound of Sugar, a grated Nutmeg; as much salt as will season it, three quarters of a pound of Beef suet shred very small, then take sweet Herbs, the most of them Marrigold, eight Spinages, shred the Herbs very small mix all well together, then take two eggs and work them up together with your hand, and make them into round Balls, and when the water boyles put them in, serve them, Rosewater, Sugar, and Butter for sauce."

They were also sugared and candied. I could keep on for many, many pages, telling of the pleasurable recipes in which this golden flower is served, but I must turn to my wine. Not only is a wine made of it, but it is used as a flavoring in other drinks, and there are recipes for both the wines and the flavoring dating from about 1600.

The Recipes

I

You need: 6 to 8 qts. marigold heads (*packed down rather firmly*)
2 lemons
4 oranges
5 lbs. sugar
2 gals. water
¾ oz. yeast (*3 packages*)

1. Gather the marigolds, using the flowers only. (You can, if you wish, strip off the calyxes, as some recipes suggest. I don't do this; I just take off the stems.) Put the flowers into a crock.
2. Wash and peel the lemons and oranges very thinly and put the rinds into the same crock. Then squeeze the citrus juice and add it.
3. Dissolve the sugar in the water, bring to a boil, and pour it over the flowers and fruit.
4. Allow it to cool while you dissolve the yeast in ½ cup warm water. Add the yeast mixture to the liquid when it is lukewarm.
5. Cover the crock and put it in a warmish place (65°-70°) to ferment. This will take from fourteen to twenty-one days.
6. Strain into glass jars and let stand for a week or two. If it is not golden clear by then, fine with eggshells or isinglass, decant, and bottle. This will make a dry, flowery wine. You can sweeten it with sugar syrup to suit your taste.

II

You need: 2 gals. water
7 lbs. sugar
1 lb. raisins
2 lemons
2 oranges

Flower Wines * **195**

> 8 qts. marigolds (flower heads only) (don't pick them all at once)
> ¾ oz. yeast (3 packages)

1. Put into an enamel pot the water, sugar (use 2 or 3 pounds more if you want a sweet wine), finely cut raisins, and the thin rinds and the juice of the well-washed lemons and oranges. Boil for about five minutes, then pour the liquid into a crock.
2. Now add 2 quarts marigold flowers and stir all well with a wooden spoon.
3. Let cool, and when lukewarm add the yeast, dissolved in ½ cup warm water.
4. Cover and put in a warm place (65°-70°).
5. The fermentation will begin in a day or two, and as soon as it starts, put in another quart of flowers, stirring well. Add a quart of flowers daily for six days, so that all 8 quarts have been used.
6. Let the must rest until fermentation stops. Clear and proceed in the usual way.

III

> You need: 4 qts. marigolds
> 1 lb. raisins
> 2½ gals. water
> 6 lbs. sugar
> 1 egg white
> 2 oranges
> 2 lemons
> ¾ oz. yeast (3 packages)

1. Gather the marigolds when the sun has dried them. Break away the flowers and put them into a crock.
2. Put in the raisins, cut up.
3. Boil the water and sugar (use 8 pounds if you want a sweeter wine) for thirty minutes, skimming off any impurities that come to the top.

4. Stir in the egg white, well beaten.
5. Pour the liquid onto the flowers and let stand for twenty-four hours, well covered.
6. At the end of that time, stir with a wooden spoon for twenty to thirty minutes, then let it stand for another twenty-four hours.
7. Strain, return to the crock, and add the thin rinds and the juice of the oranges and lemons.
8. Heat the liquid to lukewarm. Dissolve the yeast in half a cupful of the warm liquid or in warm water, and add it.
9. Let stand for a few days until fermentation stops, then strain the wine into glass jars.
10. Let these stand for a week or two, and if the wine is not brilliantly clear by then, it should be fined. Then bottle, letting the bottles stand upright for three to four weeks. Decant again and cork.

Marigold wine should not be drunk for two years.

Whichever way you make it, you will have a wine such as you have never dreamt of—golden brown in color, with a rare and unusual bouquet and a refreshing taste.

[*Pansy Wine*]

"AND THERE IS PANSIES, that's for thoughts." I wonder how many thousands of times this line of Shakespeare has been quoted!

It is difficult to make pansy wine unless you are a pansy fancier and find time to grow many of them in your garden (as I do), and unless you are a home-wine-making addict and are determined to enjoy unusual bouquets of folk wines (as I am). So here it is for your pleasure, even as it has been for mine. There are other reasons for including it in this book. The pansy has unusually lovely tales with faery glimmer all their own, worth finding and worth telling, and my own little adventures in early life are connected with them. I had so fierce a love for the multicolored velvety flower that I would rob gardens in the city to have them with me. To look at them, touch them gently, and smell their delicate scent gave me a vague thrill of mystic worlds.

The Hindus associate the scent of flowers with the five heavens—perhaps scents are my five heavens.

I have just picked a few pansies of regal purple and gleaming yellow. They are before me in a little gilded Venetian vase, and as I write these lines I touch them with my finger tips and breathe their mysterious scent. Yes, the Hindus understand the perfume of flowers, and so did Baudelaire. I try to.

There isn't a flower in existence that has been christened with more names than the pansy. The English name common today comes from the French word *pensées*, meaning "thoughts"; but it also has innumerable poetic folk names throughout the world—sixty in English and over two hundred in different lands.

It is a symbol of the Trinity because there are three different colors in a single flower. So it is called "herb Trinity" and used on Trinity Sunday. Others called it "three-faces-under-a-hood."

I first knew it under the German name, *Stiefmütterchen* (little stepmothers). In this language it is a soft, somewhat sad word. Since then, my favorite names are gay: "kiss-me-at-the-garden-gate," sometimes "behind-the-garden-gate"; "jump-up-and-kiss-me"; "tittle-my-fancy"; "kiss-me-ere-I-rise"; "pink-of-my-John"; and "cuddle-me-to-you." I wonder why, by whom, and under what circumstances these lovely names were created. There also is the well-known "heart's-ease" used by poets and rhymsters.

First I want to tell you how the fairies created the flower. There was a great to-do among them in preparation for the celebration of Midsummer Eve. Each Midsummer Eve, one of their joys was to create something to brighten and enrich the earth.

Cried one: "Let us create a new flower, a specially beautiful one." Here was the perfect gift to the world on Midsummer Eve. So they took floating blues from the sky, sumptuous reds and purples from the sunset clouds, opulent yellows from the sun, the rich browns from the earth, and painted and sang, and soon there stood the first pansy.

When I was a child, I used to watch the playing of a game and the telling of a tale with a pansy. It was the typical fairy-tale type.

Once there lived a poor little fat king who was always ailing, and most of all he had cold feet, so he kept them in a narrow tub of hot water. (Look at the small pistil of the pansy, and see how it looks like a figure, and you'll also see two feet inside a narrow, boatlike, tublike space.)

This old, sick king had a dreadful second wife and two stepdaughters. She was always nagging him for money to buy rich clothes for them all, and she went to no end of trouble to please her two daughters. (Look at the glorious three-colored petals, and find the three sepals underneath—their chairs.)

But the king's own daughters, from his first wife, had only poor clothes and but one stool to sit on. (Look at the less brilliantly colored petals, and if you tear them off, they have but one sepal, their stool.)

That was the poor king's family, and we children would play a game of this tale with the pansy. One would tell the first part of the story and tear off the rich-colored petals of mother and daughters and show them to the others. The next would tear off the other two petals, the king's daughters, and show the one sepal, the stool, to the players. And finally, there would be left the large corolla, and that would be taken away and we would all look at the fat little king, sitting in a green chair, dressed in a pale yellow coat, with his thin little legs in a narrow tub.

The juice of the pansy has long been famed in poetry. Who does not remember hearing Puck ordered to bring the juice of "love-in-idleness" (the wild pansy), which, when

> ". . . On sleeping eye-lids laid,
> Will make or man or woman madly dote
> Upon the next live creature that it sees."

And with this, let us turn to a different magical juice from this faery-colored velvety flower: its wine.

The Recipes

I

I am putting first a magnificent eighteenth-century recipe set down by a good farmer's wife of Herefordshire, England, who said, after drinking it:

"It be verrie strong and did make my head to busse and the chairs to dance so that I did go to my bed chamber and rest awhile."

And if you want your head to "busse and the chairs to dance," you make it. I did, but I did not drink enough to make my head busse and my chairs dance. The recipe is slightly modified from the original, for today's usage.

You need: *1 to 4 qts. pansies, lightly packed*
            2 lbs. sugar for every gallon of water
            1 oz. powdered ginger
            3 lemons for each gallon of water
            3 oranges for each gallon of water (if oranges are too expensive or out of season, double the number of lemons)
            3 apples for each gallon of water
            1 gal. water for each quart of pansies
            ½ oz. yeast (2 packages) for each gallon of water

1. Pick 1 to 4 quarts pansies (I use 2 quarts), and lay them in the sun to dry for three days. Be sure to keep them out of the dew and rain.
2. When the petals are dry, put a layer of them on the bottom of your crock, a good layer of sugar over that, and a slight sprinkling of powdered ginger. Then put another layer of leaves and sugar and powdered ginger (sparingly) until all the petals and sugar have been used.
3. Let stand for three days, mashing every day with a wooden spoon.
4. On the fourth day, wash and pare the lemons and oranges thinly, put the rinds into the mash, and also add the juice of the fruits. Peel good-sized apples, cut them up, and add them.
5. Heat the water and pour it into the crock.
6. Dissolve the yeast in ½ cup lukewarm water and pour it into the lukewarm liquid. Cover and let ferment until the fermentation ends (about fourteen days).
7. Strain into glass jars and add a little less than a quart of brandy if you want your head to busse and the chairs to dance. Then let the wine stand for two or three weeks. If not completely clear, fine, then bottle and cork.

Under no circumstances should you drink the wine until it is at least a year old. I have a little more to say about this drink under the chapter on brandies and unusual drinks.

II

Here is another good recipe.

You need: 1 qt. pansies, packed rather firmly
2 gals. water
2 lbs. clear honey, unblended and uncooked
2 lbs. sugar
1 lb. raisins
½ oz. yeast (2 packages)

1. Gather the pansy petals and spread them in the sun to dry.
2. When they are dry, put into an enamel pot the water, honey (orange blossom, preferably, though any other will do), and sugar. (Use more than 2 pounds if you want a very sweet wine.) Stir until the sugar and honey are thoroughly dissolved.
3. Cut up the raisins as well as you can, add them, and then put in the pansies.
4. Bring the liquid to a boil and boil it for ten minutes.
5. Pour it into a crock and let it cool off.
6. Dissolve the yeast in ½ cup warm water. Add it to the crock when the contents are lukewarm.
7. Cover and let ferment, stirring once every day.
8. At the end of a week, strain the must into an enamel pot, squeezing every drop of liquid out of the pansies, then pour it back into the crock. Let it remain there until all fermentation has stopped.
9. Strain the wine into glass jars and let it rest for two or three weeks. If not clear by then, fine with eggshells or isinglass, then decant and bottle.

The wine should not be drunk for a year.

 Rose Wine]

"THE ROSE is the honour and beauty of floures,
The Rose is the care and love of the Spring:
The Rose is the pleasure of the heavenly Pow'rs.

"The Boy of faire Venus, Cythera's Darling,
Doth wrap his head round with garlands of Roses,
When at the dances of the Graces he goes."

Thus sang Anacreon, of Teos in ancient Asia Minor, echoing the eternal praise of the rose sung long before him and since.

In a different mood, I often repeat Christopher Smart's apostrophe to the rose, from *Rejoice in the Lamb:* ". . . bless Christ Jesus with the Rose and his people, which is a nation of living sweetness." If only the poet's words would come true!

This is a difficult chapter for me to write: there is so much to say that even a whole book would not suffice. The rose is the fire opal that has burned brightest in my love for flowers since my earliest childhood. I always thirsted for sharply scented roses, and the perfume sometimes had a strangely intoxicating effect on me. And now, finally, I have a garden rich in these wondrous fair flowers. When I come out on my lawn I have roses all around me. Highly scented old-fashioned roses, and modern princely bred roses.

I hold that to truly enjoy the wines you bring into existence, you must know the rich lore of the fruits and flowers from which these wines are created. For that reason, I have gleaned and displayed a little from the immense treasury of the folk imagination that relates to these flowers and fruits. But what shall I choose from the gigantic repository the world has gathered about the rose? The tales dealing with the life, color, perfume, thorns, and lore of roses fill the world like pink soft clouds in the endless heaven. Even to give a few choice morsels of the life and scent of this Queen of Flowers is difficult. What should be chosen? Where to begin? The rose is associated with more legends, more myths, more lore and meanings, customs, romance, and poetry than any other fruit or flower. More musical lines have been written about this flower than about any other; it is on more coins, paintings, coats of arms, seals, flags, and precious *objets d'art*. It is in history, literature, religion, romance, and

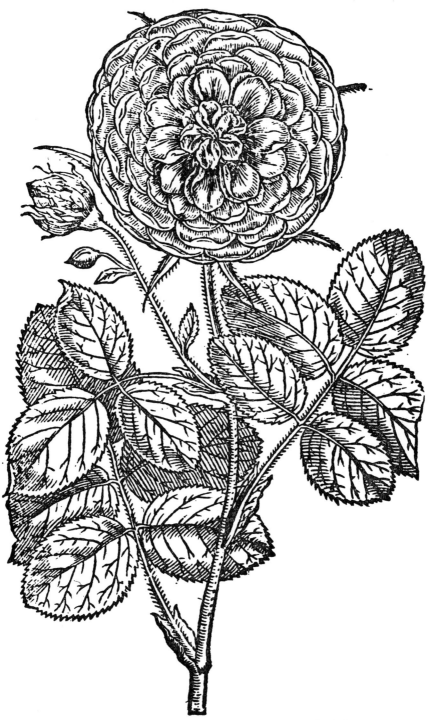

ROSE

fable. There is hardly a field of life in which roses have not had their high place, and so I shall put down only a few choice broideries from this fabric of universal beauty.

What a lovely sound is the very name of the flower: rose! It has the music of the roseate sunrise. Some say it came from the name of Eros, the God of Love. That is why nearly all the world calls the Queen of Flowers by almost the same name. That is why it is the favorite word for poet's music: rose clouds, lips soft as roses, rosy blushes, rose-fingered dawn, and so on without end. It is the flower most often connected with women. It has been part of more love potions than any other flower.

There is a deep touch of human feminine longing scented with perfume in the story of the Persian woman who made a delicately fragrant rose water to bring back her errant husband. And what of queenly Cleopatra, who strewed the floor of her palace with rose petals eighteen inches deep, to receive the man she desired, Mark Antony? There is literally a sun-gleaming avalanche of singing tales, puissant poetry, ravening myths, and high moments of the rose swirling in rose colors throughout the whole world.

Roses ran rampant in Greek and Roman mythology. It was the flower dedicated to Venus, and gods often either hold roses in their hands or, like Cupid, are crowned with them.

It was also an emblem in the insignia of cities. Wreaths of roses were on chariots, on triumphal arches, on the shrines of the great and noble. They were worn by brides and were given as prizes in competitions and used to enrich foods and wines. Rose wine was famous among the Romans. Heliogabalus not only drank it (too much of it) but bathed in it. Thousands of dollars were spent by emperors and rich men for the flower. Nero spent $150,000 for roses, one winter day, to embellish a single feast.

The Persians valued the rose and its perfume more than any other flower and perfume. Even Zoroaster, who brought the worship of the sacred fire to the earth, is connected with the rose. As in the Israelite legend, burning stakes to destroy the newborn prophet turned into a bed of roses.

What is lovelier in sentiment than the Persian poetic thought that the nightingale (bulbul) cries out in sorrow when a rose is plucked? The bird hovers around it, and when the rose blooms, drinks its perfume until it is so satiated that it drops to the ground. And when the nightingale thrills with its song, the rose bursts open to listen the better to its lover.

It would appear that poets and myth-makers vied with one another to create the most beautiful story for the creation of the most beautiful of flowers.

There is that burning tale of the birth of the rose, of the incomparably lovely Greek maiden, Rhodante, more beautiful than Helen. For only a city fought for Helen; but the Greeks elevated pearlet Rhodante to take the place of Diana, the goddess. The maiden had fled to the temple of the fleet-footed deity to escape her importunate lovers. The men attacked the temple, broke the shrine, and placed Rhodante there "as more divine." Then Phoebus Apollo, Diana's brother, torn by fury at the insult by a mortal to his sister, turned the maiden into a flower—the rose.

> "Her Arms stretch'd out are cover'd o'er with Leaves;
> Tho' chang'd into a Flower her Pomp remains,
> And lovely still, and still a queen she reigns."

This, from the Jesuit poet Rapin. And when she became a flower, she took Zephyr as a lover and would open only at his caresses.

Different lands suggest different origins of the rose. It was said to have come to the world when Venus emerged from the sea near Cyprus. Another tale relates that the flower was created by Cybele and nourished by the nectar of the gods.

The rose has more symbolic meanings than any other flower. This, from earliest times down. Incidentally, roses were probably the very earliest flowers ever drawn. I saw them in Knossos, the Minoan city uncovered in recent years on the island of Crete.

To the ancient Egyptians, and also among the Romans, the red rose symbolized joy, secrecy, and silence. Thus, in Roman times roses were suspended over guests' heads to warn them not to blabber what was heard at a feast, and they were also painted on ceilings for the same purpose. We still use the expression *sub rosa* to mean "privately" or "secretly." In the East, the rose stood for virtue and loveliness. In the Catholic Church it represented the blood of the martyrs. During the Middle Ages it symbolized life. Time changes very little, for today it is the symbol of love and beauty.

In the Bible there is mention of the rose of Sharon. And among the Israelites there was the tale of the origin of red roses—from the burning firebrands that surrounded an Israelite maiden who was unjustly condemned to be burned.

If you do not favor that version of how the rose got its color, you might prefer this one from the ancient Greeks, which relates that Cupid upset a cup of nectar in heaven; it fell on the earth and colored the growing rose.

Or that the rose turned red from blushes, or perhaps from the blood of divine Venus's foot when she ran wildly through the woods to rescue her dying lover, Adonis. Some relate that it grew from the blood of bleeding Adonis.

Attar of roses is said to have originated during the reign of the great Mogul Emperor, Jehangir. On a feast day given in honor of Princess Mour-djuhan Beygun, who was of such exquisite beauty that she was named "the sun amongst women," the Emperor, for her pleasure, filled the pools in his garden with rose water and rose leaves. The sun drew the oil from the leaves to the surface. An attendant, seeing the particles floating on the water and fearing for his head, skimmed them off, and then the oily globules burst open, giving forth a divine scent that was brought to the attention of the royal pair. This most exquisite of perfumes was named "a'ther Jehangiri"—the perfume of Jehangir. . . . Bemused and captivated, the noble Emperor murdered the exquisite lady's husband in order to marry her.

Even the thorns of the rose have received no end of attention from poets and myth creators.

Loveliest of all the tales is, as usual, the Greek. Cupid loves the rose and kisses it every time he sees it in glowing bloom. One time he kissed his love while a bee was busy in it gathering rose honey. The bee, angry at the disturbance, stung the lips of the God of Love. This angered Venus, his mother, and she put sharp stingers along Cupid's arrows. When he playfully shot the rose with one of the arrows, the stinger-thorns were set along the stem of the rose.

And there is the strangely beautiful gleaming winter tale of Saint Francis, my favorite saint. On a cold winter day when icy winds blew into his cell (I have been in it, and I know those winds on the earthen walls), a black demon came to him and whispered in his ear words and pictures of warmth and comfort. But Francis wanted to think only of helping mankind, not of ease and luxuries. So as not to think of worldly pleasures, he went out into the cold and rolled in the black rose thorns that stuck out of the snow. From the black thorns, now sprinkled with his red blood, sprang up roses of Paradise, which he offered humbly to the Lord and to the Madonna.

Saint Benedict of Subiaco grew a garden of sweet-smelling roses for their thorns—to chastise himself when sinful thoughts came to his mind.

Then there is, of course, the rosary—first begun by Saint Dominick to commemorate the hours when he was shown a chaplet of roses by the Holy Virgin. A perfect rosary has fifteen large roses or beads, representing the paternosters, and 150 small roses or beads, representing the Ave Marias. Rosaries are common in other religions: those of the Indian Buddhists have 99 beads, and those of the Chinese and Japanese Buddhists contain 108 beads, one for each possible sin in the day.

In the martyrology and in the lives of the saints there is no end of rose lore and—roses. At the beginning of Christianity, the Christians did not approve of the Queen of Flowers. They still remembered that it was a favorite flower of the Romans who had tried to "liquidate" them in the arenas. But in time, the Church could not resist the beauty of the flower or the magic of its perfume, and so the Virgin Mary became the Rose of Heaven, the Rose of Sharon, the Mystic Rose. In Rome, one Sunday belongs to the rose: the mid-Lent Sunday when the Pope blesses and embalms a golden rose adorned with jewels, praying over it and calling it "Jesus Eternal Rose that has gardened and embalmed the world." This is then bestowed on one whom the Pope wishes to honor. It was Pope Julius II who, in 1511, sent Henry VIII of England a rose plant with branches and leaves and flowers of gold set in a vase filled with gold dust instead of earth. The greatest Italian poet sang of this flower as: ". . . the rose wherein the World divine was made incarnate." Dante's last circle of Heaven blossoms as a great rose.

What is lovelier than the tale of the Hungarian Saint Elizabeth (repeated in many lands), which I heard in my childhood? The loaves of bread she was taking to the poor, against her husband's orders, turned into roses when he approached! Five roses grew from the face of Isobert, a pious monk, when he fell down dead whilst worshiping at the shrine of the Virgin Mary.

There is a medical tale from ancient Rome, of Milto, the fair maiden. She was as devout as she was beautiful, and she brought lovely flowers to the temple of Venus every day. She could afford neither gold nor silver. But alas! Soon a growth began to appear on her chin, threatening to destroy her great loveliness. Then the goddess to whom she was so devoted came to her rescue. Venus appeared to her in a dream and told her to apply to the growth roses from her altar. Milto obeyed and the growth disappeared. From then on the Romans used the rose medicinally.

As I said, throughout the ages it has been the flower most used in medicines, perfumes, beauty lotions, and poisons. It would be folly to mention the vast horde of recipes found amid the world's writings in every land. There are thousands of them. As a child, I remember a Swabian woman sitting in the market place, selling lovely flowers and yellow butter on thick green leaves, and trying to stop her nose from bleeding by putting rose petals into it. And I also remember hearing it said that putting roses on your head would take away a headache. This I think dates back to a very old belief—and I hold to it. The very thought of roses should dispel pain.

Many rose festivals have been held from the earliest days, and are still common today in every part of Europe and in North America. There is the crowning of the Rose Queen in Toulouse, France, as there is in California and Washington and other states. There are similar festivals in Italy. In Persia, where there is a continuous feast during the month of roses (when they are in bloom), there was and still exists in some places a lovely custom. When the flowers are blooming, young men pluck them in the gardens and bring them to public places; and there, with singing and dancing, they pelt the passers-by with the scented blossoms.

I mentioned our own rose festivals in North America. Let me add that we, too, have tales and customs connected with the rose. There is the lovely Chippewa Indian legend dealing with the queenly flowers and the reason for their thorns. The rose was as sweet to eat as she was beautiful to see and sweet to smell, and every creature tore it and ate it—men, animals, and insects. Then the rose prayed to Manabozho, the Great Spirit, to protect it from such destruction. He listened to the prayer and put thorns all over the stems, giving pain to those who would touch it.

A lady from Virginia told me that beautiful roses taken from "her" state (God's Own Land) into another will die of sorrow and loneliness.

Payment of rent with a rose is a custom quite common in many lands, including our own. In Mannheim, Pennsylvania, the town folks still, on a certain day each year, pay a rose to the descendants of Baron Stiegel, who gave them the land for their church.

Finally, roses have been used for thousands of years to make wines and to flavor drinks. The Persians had a secret way of making their rose wine. It was so potent that "a glassful would make the sternest man merciful, and make the sickliest mortal slumber amidst his pain."

I must note here another vinous comment. In the seventeenth century, eating dried rose petals steeped in wine was said to cure the unpleasantness of a hang-over the morning after. "It dissipated the fumes of the wine."

Theophrastus ". . . called the rose the light of the earth, the faire bushie toppe of spring, the fire of love, the lightning of the land." John Parkinson said so in 1640 in his *Theatrum Botanicvm*, and I say the same 320 years later, in my *Folk Wines*.

And on this note, with a feeling of regret, I will turn to the wine recipes.

The Recipes

Emperor Heliogabalus said rose wine helped the digestion and was so beneficial in every way that he filled the public baths with it.

I thought of this as I looked at my three-gallon jars of clearing rose wine and decided it was the proper moment to set down how to make it. This is how I made the wine that is clearing in the jars, and how you can make it too.

I will begin with an ancient yet practical recipe by Dioscorides; slightly modified, it can be used today.

I

A rose wine

"Throw 5 pounds rose petals, picked the day before, into 10 sextari [11.2 liquid pints] of old wine, and add, at the end of thirty days, 10 pounds clear honey."

I am certain it will be a splendid rose-scented wine. You need not follow the exact proportions. Last year I threw a fistful of scented rose petals into a pint of brandy, and then, after three months, drank a rose-scented and rose-flavored brandy.

II

Now to a modern recipe.

You need: *4 to 8 qts. rose petals, firmly packed down*
2 gals. water
5 to 8 lbs. sugar
2 lemons
½ oz. yeast (2 packages)

1. Pluck enough scented rose petals to fill 4 to 8 quarts, and let cold water run over them for about ten minutes to wash off any spray dust. It is best to pluck the petals at least ten days to two weeks after the roses were last sprayed. Put the petals into a crock.
2. Put the water into an enamel pot and bring it to a boil. Put into the water 5 to 8 pounds sugar, depending on how sweet you want the wine to be. When the water boils, pour it over the rose petals in the crock.
3. Peel the washed lemons very thinly and put the rinds into the crock, then add the lemon juice.
4. Dissolve the yeast in ½ cup warm water and pour that into the crock.
5. Let the must ferment, stirring it once or twice each day.
6. When fermentation stops, strain the wine into glass jars, squeezing every drop of juice out of the rose petals. Let it stand for two or three weeks, and then siphon off the clear wine into bottles. If the remaining wine does not clear of its own accord, fine it with eggshells or isinglass.

I wanted my rose-petal wine a little sweeter than it turned out, so I took a quart of the wine, put it into an enamel pot, and added 1½ pounds sugar. I heated this slowly until it began to simmer. I let it simmer for about fifteen minutes and then poured it back into the rest of the wine, which made just the right sweetness. I let the wine stand in the gallon jars for four more weeks and then decanted it into bottles.

Of course, with each decanting a little of the wine is lost; but it is better to have less wine, without the taste of the lees, than more with a harsh flavor.

III

You need: *4 to 8 qts. rose petals, firmly packed down*
 2 gals. water
 6 to 8 lbs. sugar
 ½ oz. yeast (2 packages)

1. Follow step 1 of the preceding recipe.
2. Bring the water to a boil and pour it over the petals.
3. When the water has cooled, squeeze the petals by hand, again and again. I did this almost every hour for a whole day.
4. Strain into an enamel pot, squeezing every possible drop of juice from the petals. Then add the sugar, bring to a boil, and let simmer for about twenty-five minutes.
5. Pour back into the crock. Dissolve the yeast in ½ cup of warm water and put that in.
6. Cover and let ferment. When the fermentation is finished, proceed to clear as in the previous recipe.

Don't drink the wine for at least one year, then serve it cold. It will always give rise to much discussion and create much delightful pleasure.

CEREAL & VEGETABLE WINES

 Barley Wine]

BARLEY WAS CULTIVATED from time immemorial and has often been called the most important cereal of the world. It has always grown everywhere.

Among the remote Greeks, it was held in great esteem. The first part of the name of Demeter, the ancient deity, is presumably derived from a Cretan word meaning barley. Thus, Demeter is the barley god, or "Barley Mother." Barley was also one of the most common Cretan foods.

The same held true among the ancient Aryans.

In India, barley was the symbol of abundance. The name of the god Indra means "he who ripens barley." Naturally this cereal held a high place in many Indian ceremonies: births, marriages, sacrificial rites, and funerals. All peoples used this grain in their religious ceremonies.

During my early life in Austria, when I spent the summers with my grandfather, who lived in the small village of Koroluvka, I saw many gay festivals at the time of the barley harvest.

There were crowds of peasants and their children, dressed in their white and multicolored clothing—reds, blues, yellows; all the colors of the rainbow—embroidered in intricate designs. Some young men played on willow flutes, with sometimes a fiddle or an accordion as accompaniment. One red-faced girl wore around her head a wreath of barley stalks; she was the queen of the festival. I have learned since that in many lands it is the custom to crown a maiden with the last stalk of grain gathered. She is the Barley Mother for the year. I came across the same custom while collecting folktales in Latin America.

There are innumerable primordial ceremonies and rites, dark and moon-colored, connected with these mothers of the fruits of the earth, fitting into the stark pattern of early folk life.

In Scotland, a figure called Carlin was made from the last grains and stalks gathered. In Poland, the figure was called Baba (the old woman). When the figure was completed, it was carried with songs and dances to the largest, most important farm.

Barley has been used in drugs and beautifiers: ". . . you strain Barley water with a few drops of Balm of Gilead," and this, when shaken constantly for ten to twelve hours, would give a lotion that ". . . marvelously embellishes a complexion and takes away wrinkles. It should be used every day for a long time."

In the Far East, barley was used for similar purposes. There is an ancient Persian recipe for a beauty bath for which barley was prescribed.

Finally, the fermentation and distillation of barley make a potent wine. I am now using barley wine to make my own cordials. If you let your barley wine rest for a few years, you will have a drink more potent than sherry. Under no circumstances should it be drunk young. It must be at least two years old, and even then it will still be somewhat raw.

The Recipes

I

The simplest and best recipe I know is the following:

You need: *2 lbs. barley, coarse, if possible*
2 lbs. potatoes—you can use 4 lbs. barley and eliminate the potatoes
2 oranges
2 lemons
2 lbs. raisins
2 gals. water
8 lbs. sugar
½ oz. yeast (2 packages)

1. Try to find barley used for feeding fowl; if you can't get that, get cooking barley and put it into a crock. Also 2 pounds of old potatoes, well washed and cut into quarters (or use 4 pounds of barley and no potatoes). Put in the thinly pared orange and lemon rinds. Cut up the raisins as small as you can and put them into the crock.
2. Put the water in an enamel pot and bring to a boil.
3. Dissolve the sugar thoroughly in the water.
4. Squeeze the juice from the lemons and oranges and add it to the liquid. Let it cool and put it into the crock.
5. Dissolve the yeast in ½ cup warm water and pour it into the lukewarm liquid. Cover it well and stand it in a warm place (65°-70°). Fermentation will take about twenty-one days. Strain, fine and rack if necessary, siphon, and bottle.

II

Here is another way of making barley wine during the high summer months when the bright flowers are in bloom.

Since barley wine, by itself, has just a strong liquor taste, you can flavor it with flowers to suit your inclination. You can use rose petals or marigolds, which are to be found in abundance during these months.

Eliminate the potatoes, and follow the previous recipe exactly, with the following additions. For each gallon of wine, take 2 quarts ripe and freshly cut rose petals with the white ends cut off, or the same amount of marigold flowers, and add these to the contents of your crock.

Ferment for two weeks. Strain the wine and let the fermentation continue until it is finished. Then follow the regular procedure.

Your barley wine will be a strong wine, particularly if you let it age.

[Beetroot Wine]

DEEP RED BEETROOTS, notwithstanding their royal color, have little lore about them, but they make an excellent red wine, pleasant to taste and pleasurable to see.

In taste, it measures up to a light port, particularly if given time to mature: from two to five years. The simplest procedure is to make two to four gallons every year and keep on putting half of it away. This is a good practice with all wines.

The Recipes

I

You need: 2¼ gals. water
6 lbs. beets
2 lemons
12 cloves
½ to 1 oz. fresh ginger root
2 sticks cinnamon
5 peppercorns
1 lb. barley (*optional*)
½ oz. yeast (2 packages)

1. Put the water into an enamel pot.
2. Wash the beets, bought or home-grown, cut them into small pieces, and put them into the water.
3. Wash the lemons well. Cut off the rinds very thinly and put them into the pot.
4. Make a small bag of any white material and put the spices in it. Tie the bag with a long string and hang it into the pot so that it is immersed in the liquid. Bring to a boil and let it boil for at least thirty minutes or more, until the beets turn white. You can, if you wish, take the spice bag out then; or, for a spicier wine, you can let it stay in the must during the fermentation period.
5. Strain the boiled liquid into a crock, squeezing the beets well through the cloth. You may add a pound of barley to the crock if you wish.
6. Dissolve the yeast in ½ cup lukewarm water and put it into the liquid when it is lukewarm. Cover the crock and put it into a warmish place (65°-70°) to ferment. This will take from fourteen to twenty-one days. When the fermentation has stopped, strain, clear, fine, siphon, etc., until the wine is perfectly ruby-

colored. At this point you may add a pint of brandy or of barley wine or potato wine to strengthen it, but this is not necessary. Bottle and cork the wine.

It is excellent only after it has been given time to mature (at least one year).

II

This is a slightly different way of making beetroot wine, following an old recipe.

You proceed just as in the first recipe, but after boiling and straining the beets (be sure to take out the spice bag), you add a pound of parsnips, cut into small pieces. Then boil the liquid again for thirty minutes. Since parsnips, notwithstanding their prosaic name, have a fine, exotic flavor, they will impart to the wine a rare and delicate Eastern savor that is delightful—a new gustatory experience.

III

Here is a wine made from beet leaves.

You need: *2 gals. water*
2 lbs. large, full-grown beetroot leaves (2 to 3 full quart jars)
2 oranges, washed and sliced
2 lemons, washed and sliced
spices (as in previous recipe)
8 lbs. sugar
½ oz. yeast (2 packages)

1. Put the water into an enamel pot and add the well-washed beetroot leaves, the oranges and lemons, and the bag of spices.
2. Bring the liquid to a boil and let it boil for about thirty minutes or a little less.
3. Add the sugar, dissolve it thoroughly, and pour all into a crock.

4. While it is cooling off, dissolve the yeast in ½ cup lukewarm water and add it to the liquid when it is lukewarm.
5. Cover, put in a warm place (65°-70°), and let it ferment. When the fermentation has stopped, strain, clear, and fine, and then bottle and cork. This wine also needs long maturing. You can vary it by adding dates before the boiling. This will alter the flavor of the wine and add to its piquancy. Or, you can impart an orange flavor, or a heavy "brown" flavor, to beetroot wine by adding half a dozen oranges or a pound of prunes. Whichsoever variation you make, you will get an excellent after-dinner drink.

 Carrot Wine]

THE MORE ONE LEARNS of the varied and wondrous lives that the products of the earth have led through the centuries, the more fascinated one becomes. And one of the final wonders is the transformation of these products into stimulating wines.

Take that simple and excellent nutritious fruit of the earth— the carrot. (I don't like the word "vegetable." It has a thick, unsavory sound that is unpleasant to the ear.)

In ancient times, the Greeks used it as a medicine to help those in love, and it was called *philtron*. Carrots were always used in their love philters to inflame desire in men and women.

Centuries later, lo and behold! The solid Dutch burghers forbade the public sale of carrots because the lovely, bright, red-orange color excited folks too much and reminded them of the House of Orange, which had fallen into abysmal disgrace.

On the other hand, a favorite dish in England around 1720 was a "carrot sippet." Carrots were boiled and pounded; butter, white wine, salt, cinnamon, and shredded dates were added, and the mixture stewed. It was served on toast or in a dish garnished with hard-boiled eggs cut into halves or quarters and sugared. This made a sweet accompaniment to meat.

Here is a simple, delectable way of serving carrots, created in mine own domain.

Boil whole carrots until they are fairly soft, then put them into a flat glass dish or pan, sprinkle them well with cinnamon, and cover them generously with mead. Let them simmer until ready to serve —say about half an hour. They are delicious. I speak from eating knowledge. And I speak from drinking knowledge when I tell you that the following recipes make excellent wines.

The Good Carrot Wine

I

You need: 2¼ gals. water
          7 lbs. sugar
          10 lbs. carrots
          4 oranges
          2 lemons
          1 lb. raisins
          ½ oz. yeast (2 packages)

1. Put the water into an enamel pot. Dissolve the sugar in it.
2. Wash the carrots but do not peel them. Try to get sweet carrots. You can find out if they are sweet by tasting them. Add the carrots to the water and bring to a boil. Let it simmer until the carrots are soft.
3. Strain the liquid into a crock. You can eat the carrots.

4. Wash the oranges well, slice them, and put them into the liquid. Add the lemon rinds, pared very thinly, and the lemon juice.
5. Cut up the raisins a little and add them. Let the liquid stand until it cools off.
6. Dissolve the yeast in ½ cup warm water, and add it to the liquid when it is lukewarm. Cover and let stand in a warm place (65°-70°). Fermentation will take from fourteen to twenty-one days.
7. Strain the juice into jars and proceed to clear by fining until it is a perfectly brilliant, golden color—as mine always is.

II

Here is another way to make carrot wine. It does not vary very much from the preceding, but it will give a somewhat heavier wine.

You need: *10 lbs. carrots*
2¼ gals. water
1 oz. fresh ginger root
7 lbs. sugar
4 oranges
2 lemons
2 lbs. raisins
½ oz. yeast (2 packages)

1. Wash the carrots thoroughly, cut them into small pieces, and put them in a muslin bag. Put this into an enamel vessel containing the water.
2. Add the ginger root.
3. Bring the water to a boil and let it simmer until the carrots are soft and pulpy.
4. Squeeze the carrot pulp until all the liquid has been expressed. Take the ginger out and put the liquid into a crock.
5. Dissolve the sugar thoroughly in the liquid.
6. Wash the oranges and lemons well. Slice the oranges and put

them into the crock. Then pare the lemons thinly, and add the rinds. Squeeze the lemons and add their juice.
7. Cut up the raisins a little and add them to the liquid.
8. Dissolve the yeast in ½ cup warm water. When the liquid is lukewarm, pour the yeast into it. Cover, put in a warm place (65°-70°), and let it ferment.
9. When fermentation has ended, strain, clear, and fine. When the wine is brilliantly clear, bottle and cork it.

Whichever recipe you use, you will have a delightful drink.

Celery Wine]

A MOST EXCELLENT WINE can be made of the white parts of celery stalks, almost in the same manner that you make carrot wine. I am fond of the celery flavor, so I make a good supply of the wine. When it is clear and dry, a glass or two is a fine accompaniment to any kind of fish.

It is said to be healthy, too—and a specific for rheumatism, when taken over a long period of time.

I can highly recommend the wine and the stalks as a flavoring in cooking—in soups and, particularly, with steamed clams.

The wine can easily be made in a city apartment. Use only the white parts of the stalks, not the green or the leaves; they will destroy the flavor of the wine.

This is the recipe.

ONION

You need: 2¼ gals. water
8 lbs. celery (white parts only)
2 lemons
2 oranges
½ lb. raisins
6 lbs. sugar
¾ oz. yeast (3 packages)

1. Put the water in an enamel pot.
2. Cut the white celery stalks into 2-inch pieces; put these into the water.
3. Wash the citrus fruit. Peel the rinds thinly and put them in.
4. Cut up the raisins and add them to the mixture.
5. Bring the liquid to a boil and let it simmer for a long time—until the celery is very soft.
6. Strain into a crock. (You can keep the celery to use as a vegetable. It has a pleasing taste.)
7. Dissolve the sugar completely in the hot liquid. Add the juice from the lemons and oranges.
8. Dissolve the yeast in ½ cup lukewarm water and add it when the liquid has become tepid.
9. Cover and let it ferment for two to three weeks.
10. When the fermentation is over, strain, fine if necessary, decant, and bottle.

Some recipes are improved by the addition of spices. I don't think they are helpful in this wine. The celery flavor is very delicate and will be destroyed or distorted by strong spices.

Do not drink the wine for at least a year.

 Onion Wine]

DON'T TURN YOUR NOSE, or eyes, disdainfully from the onion, for a good and potent wine can be made from it. Remember, it has been cultivated, and actually worshiped, through countless silent years. The Egyptians were the first to worship it, and used it as a sacrificial offering and as one of their most important sustaining foods. The great strength of the men who sweated and worked on the ageless pyramids was attributed largely to the onions they ate, and they ate very large amounts indeed.

There is an inscription in the Great Pyramid of Cheops, *c.* 2900 B.C., that tells that 1600 talents of silver were spent for onions, radishes, and garlic for the workers on the stone tomb—$3,481,600!

Many poets used these facts for rhymes. Juvenal, the incomparable satirist, found his own word for the scent and flavor of the onion:

> "How Egypt, mad with superstition grown,
>   Makes Gods of monsters, but too well is known:
> 'Tis mortal sin an Onion to devour,
>   Each clove of Garlic hath a sacred power;
> Religious nations sure, and blest abodes,
>   When every garden is o'errun with Gods."

And again:

> "Such savoury deities must sure be good,
>   It serves at once for worship and for food."

To the Hindus, too, the onion was a symbol of religious mysteries, solemnities, and divination. There are also innumerable references to it in the Bible.

Later on, the onion changed considerably historically and folkloristically. It vacillated between favor and disfavor, but folks always remembered that it had once been an object of worship.

The Arabs and the Chinese used onions to ward off witches and demons, for these evil spirits had both fear and respect for the bulbous plant. Later, in some parts of these countries, onions were thrown at a newly married couple to keep away the evil eye.

Among the Greeks, the onion was presented as one of the gifts to a newly married couple, for the same reason that we throw rice at newlyweds. When Iphicrates married the daughter of King Cotys, among the wedding gifts there was a jar of snow, a jar of lentils, and a jar of onions.

And yet, necromancers say it is a plant of ill omen. To dream of onions forebodes coming trouble. Like so many other plants, the onion has long been used on the Continent for divination and fortunetelling. In parts of England, an onion named after Saint Thomas is peeled and wrapped in a clean kerchief and placed under the pillow, then a little rhyme is repeated:

"Good Saint Thomas, do me right,
  And let my true love come tonight."

It came in for weather divination, too. There is a folk saying:

"Onion skin, very thin,
  Mild winter's coming in;
Onion skin, thick and tough,
  Coming winter, cold and rough."

The onion has had an important place in medicine from the

very earliest times, and there are many references to prescriptions in which it was used.

"The juice of the onyon poured or snuffed up in the nose-thrilles causeth one to sneeze and purgeth marvelously the brayne."

"If dropped into the ears it is a cure against deafness and humming noises and ringing—of the same."

I somewhat remember that when I was a child it was used for that purpose in Austria. And to top it all: It also "filleth agayne with hair the pyled [bald] places of head being laid thereon in the Sonne." There is a golden opportunity for the bald-headed!

Some North American Indians put onions into the sickroom to draw out the sickness. They made a syrup of chopped onions to cure colds, and they also put the heart of the onion into the ear to cure earaches. Altogether magical powers were attributed to it.

With the years, onions gained a very important place in the culinary world. There is a classic precedent for its popularity. When the goddess Latona, the mother of Apollo and Diana, once lost her appetite, she regained it by eating an onion, and for that reason adopted the vegetable, which was consecrated to her.

In the last hundred years, our own *chefs de cuisine* have learned that it enhances and sweetens the taste of good soups and is invaluable for seasoning endless dishes. Fully 80 percent of our prepared foods contain onions.

With so important a place in the culinary realm, it surely deserves a place in viniculture. It makes a strong wine with a most unusual flavor, worth cultivating.

The Recipe

I found the recipe under amusing circumstances. Liking to boast, I said to a woman in an antique shop who was showing me some beautiful old American goblets:

"This summer I made nine different fruit and flower wines."

A stoutish, simply dressed woman who was also in the shop piped up: "Bet you never made wine from onions!"

"Didn't know I could," was my reply.

"You sure can!" And then she gave me the recipe.

You need: *1 lb. onions*
1 lb. uncooked barley or potatoes
2 lbs. raisins
2 gals. water
4 lbs. sugar
½ oz. yeast (2 packages)

1. Peel and slice the onions and put them into a crock.
2. Clean and add either barley (my own preference) or potatoes, washed and sliced.
3. Cut up the raisins and put them in.
4. Warm the water, in which the sugar has been dissolved, and pour it into the crock.
5. Dissolve the yeast in ½ cup warm water and pour it into the liquid.
6. Cover, set in a warm place (65°-70°), and let ferment, stirring daily. The fermentation will take about fourteen days or more.
7. Then strain, clear, fine if necessary, and bottle.

The result will be a strong, dry wine with a most unusual taste. No, there will not be an onion bouquet!

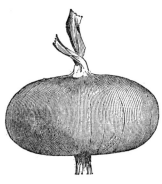

 *Parsnip Wine*]

IT MAY SEEM UNUSUAL to begin a chapter in a wine book with a cooking recipe; but eating and drinking are, and should be, so intertwined that this should be accepted as normal. Besides, it is so delightful and unusual a dish—I have never met it in any cookbook or restaurant, though I have eaten in legions of restaurants and have gone through as many cookbooks—that I think it is well worth recording for others. It was created by my lady of the house, who is a rare *chef de cuisine*. Here is how "parsnips au Melita" are made.

Boil as many parsnips as you want until they are comfortably soft. Peel them and cut them in half. Now take a flat cooking dish (Pyrex is good) and spread a bit of corn oil on the bottom. Lay the parsnips in the dish and sprinkle generously with cinnamon. Then pour into the dish a good mead (honey wine), covering the parsnips. Put the dish into the oven at low heat and let simmer very gently until a good deal of the mead has evaporated.

I promise you, from many experiences, that this dish will give you a most rare culinary pleasure.

As for the wine of the parsnip, I was greatly astonished to learn that it was one of the most popular vegetable wines in England. Probably the most important reason is that it makes a superb semisweet drink when allowed to mature for a few years.

If you grow parsnips, as I do, dig them out after the first frost. The cold adds an unusually rare flavor.

The Recipes

I

You need: *8 lbs. parsnips per gallon of water*
2½ gals. water
6 to 8 lbs. sugar
1 slice toast
½ oz. yeast (2 packages)

1. Wash the parsnips but do not scrub them; cut them in halves or quarters. Put them into an enamel vessel, laying them down flat.
2. Pour the water over them and bring slowly to a boil; simmer very gently for a full two hours. It is important that they do not become mushy, for this will bring a cloudiness to the wine that cannot be cleared up.
3. Strain the liquid into a crock and dissolve the sugar in it. (Use 8 pounds if you want a very sweet wine.)
4. Let cool until lukewarm. While it is cooling, prepare a slice of toast and cover with the yeast, dissolved in ½ cup warm water. Put them into the crock.
5. Cover and let stand in a warmish place (65°-70°) until the fermentation has stopped. This will take from two to three weeks.
6. Strain off into jars and let stand for two weeks. The wine will probably be completely clear by then. Pour off the clear wine; fine, decant the rest, and bottle.

This wine should never be drunk until it is at least two years old. The older the wine, the finer the bouquet and flavor.

II

Here is a slight variation that will give you a different-flavored wine.

Use the same amount of parsnips and the same amount of water, but add to it, before boiling: the thin rinds of two well-washed lemons and two oranges; and ½ ounce bruised fresh ginger root, or a touch of some spice you favor. Simmer this for a full two hours. Then take out the parsnips carefully and strain off the liquid into a crock. Now add the juice of the citrus fruits and let cool. While it is cooling, prepare ¾ ounce yeast (3 packages) and add it to the liquid when it is lukewarm. Then follow the usual procedure of clearing and bottling.

But keep in mind that this wine needs time to mature: at least one year, preferably two.

 Potato Wine]

FRANK ZUBEK, a Czechoslovak, worked for me for many years as a carpenter, builder, mason, and general handyman. He was six feet tall, wiry, with the strength of a bison. He was a true artist with a keen mind, one who loved his work with his heart, his senses, and his hands, as any artist must do to deserve the name. He was also a splendid eater and an even more splendid drinker. During our American Black Era, Prohibition, he brewed a potato whisky-wine in the basement of our brownstone house in that once fine old Murray Hill section of New York City, which gave him or anyone else who drank it the strength of Samson and Hercules.

One night, after drinking a goodly quantity of goodly potato whisky-wine, he began to sing and to harangue the world. The folks who lived across the yard objected to his singing and called the police. They also shouted across the fence for silence. Whereupon Frank felt he should inform them that this was America and that he was an American and had the right to sing and talk when he wished to do so. It took four burly policemen to hold him down to prevent his demolishing everything in sight. Potato wine is a strong wine—even the one I make.

It was Frank who told me that in Czechoslovakia and Hungary, during the potato harvest, the man who dug out the last potato was always treated and given plenty to drink.

Interestingly enough, this custom is also followed in Peru and some other Latin American countries, whence the potato originally went to Europe.

But in Germany, where demons run rampant and wild-eyed, they accuse the nourishing fruit of having a demon or a wolf in it—a *Kartoffelwolf*. When the last potato is dug up, the peasants dress a puppet grotesquely and carry it in procession to the house of the one who owns the potato field, with incantations and doggerel, apparently to either appease or exorcise the demon.

Quite different from the Americas! There they believe that any plant of use to life is animated by a divine spirit that caused its growth. That spirit is named "Mother." Thus there is a Potato Mother. Images made from either the fruit or the leaves are dedicated to these divine and miraculous Mothers. The images are dressed in women's clothing and are worshiped.

The potato was brought to southern Europe by some of Pizarro's *padres*. Seamen introduced it to England, and it quickly grew in popularity everywhere. At first it was considered an aphrodisiac. At one time it cost as much as $1,000 a pound. But the Scots forbade its sale on the ground that it was unholy because it was not mentioned in the Bible.

Many customs grew up around the potato, as it increased in popularity. In Ireland it was planted on starry nights and on Mondays and Thursdays to secure good crops. In some other regions, the dark of the moon was the proper time for planting. To plant on Good Friday meant a poor crop. When the first tender tubers were eaten, all the family had to join in the feast, otherwise the spirits would be offended and the crops would rot.

Potatoes were said to have curative powers against rheumatism and sciatica. This was believed in Holland, even as it was in England. In the former land, the potato had to be stolen to be an effective cure. In Devonshire, England, keeping a potato in your pocket would keep toothache away.

Warts were cured in our own good land by the tubers. Rub a wart with a potato, then hide or bury the tuber, and the wart will go with it. In Texas, scraped potatoes will cure frostbite or burns.

When I was a youngster living in Austria, I remember that a sliced potato was put on a festering skin sore to draw out the pus. We children would put chunks of raw potatoes at the end of long, thick twigs and throw them like boladeros.

Nor has the good vegetable escaped the realm of black superstition. In the southern lands, in Italy, particularly in Sicily, you could kill an enemy by writing his name on a slip of paper and then fastening it to a potato with as many pins as possible. Death would come in a month, with great pain. This, of course, is reminiscent of revenge-bringing poppets believed in throughout the world.

Potatoes, like other fruits and vegetables, are often given strange names in both our land and in England. Often they are named after the locality where they grow—Maine or Idaho potatoes. Then there are leatherjacks, Murphies, blue-eyes, flukes, Irish apples, bog oranges, Mrs. Murphy in a sealskin coat, and so on and on.

And so, from seal coats to wine. I have made it, and it has a larger percentage of alcohol than my other wines, judging by my taste buds and "inner sensation." Sometimes I use it for my fruit brandies, with good results.

The Recipes

Potato wine can be made with or without yeast. This wine, or any other wine to which potatoes are added, will have what I would call a slightly earthy taste. You can easily eliminate this by adding a few cracked pits of plums, cherries, or peaches, or a little bruised ginger. The addition of a cereal like wheat also will help change its flavor.

I

You need: 2 lbs. potatoes, medium-sized
3 lbs. raisins
2 lemons
2 oranges
2 gals. water
5 to 6 lbs. sugar
½ oz. yeast (2 packages)

1. Wash the potatoes thoroughly—do not peel them—and put them into a crock.
2. Cut up the raisins and add them to the potatoes.
3. Add the lemons and oranges, sliced.
4. Heat the water, in which you have dissolved 5 to 6 pounds sugar, and pour this over the potatoes and fruit.
5. Dissolve the yeast in ½ cup warm water and put that into the crock. Cover and let the must ferment, stirring it each day but taking care not to bruise the potatoes.
6. When fermentation stops, strain the wine into glass jars. It is at this point that you put in the fruit pits to make the flavor more pleasing to your taste. Let the wine stand for two weeks, then siphon off the clear part into bottles. The rest—or all, if necessary—you clear with eggshells or isinglass. When completely

clear, and before bottling, you may sweeten your wine with sugar syrup, spiced or pure, to create any flavor you like.

Do not drink the wine for a year.

II

You need: 2 gals. water
3 lbs. potatoes
1 lb. wheat
6 lbs. sugar
2 lemons
2 oranges
1 oz. fresh ginger root
½ oz. yeast (2 packages)

1. Measure the water into an enamel pot and add to it the potatoes, well washed and cut into halves, and the wheat grain. (If you cannot get wheat, you can substitute four shredded wheat biscuits.) Bring slowly to a boil. Boil for twenty-five minutes, taking care not to mash the potatoes. At the end of that time, strain into a crock without mashing the potatoes.
2. Dissolve the sugar in the liquid. Then add the thin, washed rinds of the lemons and oranges and their juice. Also add the bruised ginger. Replace in the enamel vessel, bring to a boil, simmer for fifteen minutes, then pour back into the crock.
3. When the liquid is lukewarm, dissolve the yeast in ½ cup warm water and put it in, then cover and let stand in a warm place (65°-70°) until fermentation ceases. Strain into glass jars and let the wine stand for two weeks.
4. If it is not clear, fine. If partially clear, siphon off the clear part and fine the rest. When it is completely clear, bottle.

This wine should not be drunk for one year. It will be a strong wine, high in alcoholic content.

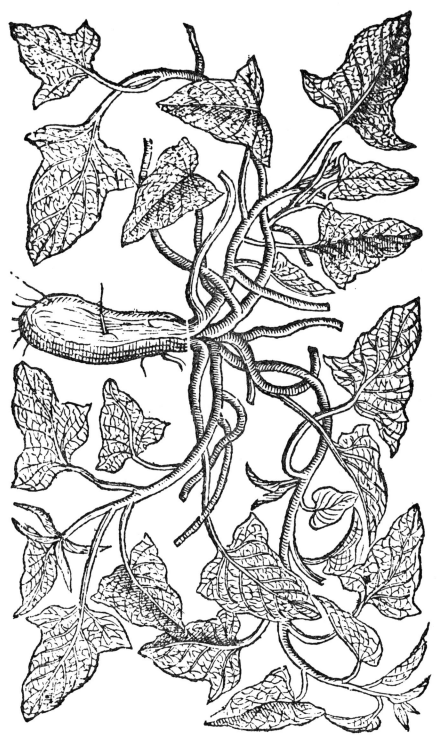

POTATO

[*Rice Wine (Sake)*]

"THE GREATER PART of the inhabitants of the Province of Cathay drink a sort of wine made from rice mixed with a variety of spices and drugs. This beverage, or wine, as it may be termed, is so good and well flavored that they do not wish for a better. It is clear bright and pleasant to the taste, and being [made] very hot, has the quality of inebriating sooner than any other."

So wrote Marco Polo hundreds of years ago, and so also say I when my friends and I sip golden-colored hot rice wine from delicate, cool, green-and-moss-gray jade cups. For I had decided that my homemade rice wine merited, and would be more enjoyable if poured from an old, lovely, carved jade bottle and drunk from soft-toned jade cups. How I finally acquired my treasures I have already related.

The Chinese wrote of rice five thousand years ago, and some say it was the first fruit of the earth that man planted when he crawled out of his primordial shell. Endless tales, most of them delicate, gossamer veils of baby's-breath, tell of the tiny, nourishing grain. Omaterasu Omu-Kanu, the Japanese Sun Goddess and ancestress of the imperial family, first sowed rice in the field of Heaven. She celebrated the feast of the "first fruits" by giving the emperor's grandson rice sprouts from the sacred heavenly garden of plants.

There is still another tale from the land of delicate art, telling how rice came to the people. Many years ago there lived a priest whose only foods were roots and herbs. One day, while sitting in silent meditation, he saw a small gray mouse bringing some rice and hiding it in the corner of its home.

"How did the little animal get this grain? From where did it bring it? Perhaps we, too, could eat it," ran through his mind.

He set a trap and caught the tiny mouse, then he tied a silken string to its thin little leg and let it run freely, holding onto the end of the silken string. The little beast scampered off, and he followed. The nimble rodent ran a long time, the priest after it, until they came to a land where rice grew as far as the eye could see. There the priest learned how to cultivate rice and brought the knowledge back to his land. It proved such a great blessing that thenceforth the mouse, under the name of Daikoku-Sama, was worshiped for the benison it had brought to the people.

The Moslems say that rice grew from a drop of perspiration that fell from Mohammed in Paradise. Thus it became a sacred food among the Mohammedans. Their national dish, couscous, was revealed to the prophet by the Archangel Gabriel.

Rice is the staple food in the East, even as wheat is in the West. In the islands of Indonesia, it is treated as if it were a pulsating human being with an individual's soul. There exists the "Rice Mother" —the spirit in the grain. When the rice begins to grow, it is treated with the care and consideration bestowed on a pregnant woman. All loud noises are prohibited in the rice fields. The first-formed rice ears are fed with rice pap. The words spoken in the fields must be courteous and gentle so as not to disturb or frighten the Rice Mother.

When the rice is to be cut, the knife is so held in the hand that it cannot be seen by the stalks, and only very special gentle language is used, so as not to frighten them.

There are, besides, all kinds of age-old emotional ceremonies when the grain is harvested. There was, and is, a mystic bride-and-groom arrangement and other acts symbolic of the abundance of life and generation, all coming from the dim dreaminess of time.

In some of the great islands, rice is worshiped as a divinity, and in Siam there is a gracious custom of offering honey to the stalks before cutting them.

In India rice is a requisite in marriage ceremonies, and it is used as an offering to the gods in many ceremonies and sacrifices, even as it is used to detect evil spirits. The custom of throwing rice at a bride after the wedding ceremonies has come down to us from India. In medieval times, it was used in beauty recipes.

As for the wine that the Chinese have drunk for thousands of years, it has been celebrated in poetry, tale, and song for endless centuries. It was a favorite theme of the immortal Li Po, who was a good drinker of rich rice wine. Read "The Eight Immortals of the Wine Cup" by this great poet.

> ". . . He drowsed in the wine shop
>   on the city street of Chang-an;
> And though the Emperor calls
>   will not board the Imperial Barge . . ."

Nearly every one of Li Po's poems is scented with the perfume of rice wine. One of his friends said, "As for Li Po, give him a jugful, he will write one hundred poems." He died with a wine cup in his hand.

And so to the recipes, to give you, if not a jugful, then a few bottles full.

The Recipes

Rice wine is quite popular in the United States of North America (as Latin Americans insist we should say, and justly so), judging from the number of recipes I have received from different parts of the country. All are very much alike. The one I am giving here is a compound of recipes sent to me by Marion Duhrels of New York, Mrs. B. Keysner of West Virginia (who was given it by Mrs. I. G.

Cereal and Vegetable Wines * 241

Gridley of Pennsylvania), and my own. All are similar in quantity and procedure.

The recipe is for 3 gallons. It is worth making this amount, for it has a fine taste and flavor when served hot. Here, too, the older the wine, the better the taste.

I

You need: *3 gals. water*
5 to 8 lbs. sugar
5 lbs. rice, preferably brown (unhusked)
3 to 5 oranges
2 to 3 lbs. raisins
¾ oz. yeast (3 packages)

1. Dissolve the sugar in the water and bring to a boil.
2. Put the rice into a crock and pour the boiling syrup over it.
3. Slice the oranges and add them to the liquid: peel, pits, and all.
4. Cut up the raisins finely and put those in.
5. Dissolve the yeast in ½ cup lukewarm water and pour it in when the liquid in the crock is lukewarm.
6. Allow this to ferment for a full three weeks, then strain the wine into glass jars and let it rest for two weeks.
7. Siphon off the clear wine into bottles and fine the rest. Let it stand until golden clear, and then bottle. Wait a year before drinking it; the older, the better.

II

When I was in Hong Kong, I learned of an interesting "ancient" recipe for making rice wine. I don't know how authentic it is, but it is a good recipe, one worth a trial. I am giving it exactly as it was told to me.

1. Let 5 pounds of rice stay in water (say 2 gallons) for three weeks, stirring it every day.

2. At the end of that time, boil, then simmer until the rice is fairly soft. Strain the liquid into a crock.
3. Now add fruits or flowers to suit your own taste. (That, of course, will depend on the time when you make it. In the summertime I would add 10 to 20 pounds peaches, or a quart of marigolds or rose petals, or one or two fistfuls of any herb. In the wintertime you could add pineapple or any of the citrus fruits.) In old China they used lemon flowers, orange juice and rinds, and other fruits and flowers.
4. Add honey to sweeten it according to your own taste. Now, when it has the flavor you desire, yeast is added and it is allowed to ferment.
5. When the fermentation is finished, strain the wine into glass jars or gallons, allow it to stand, and then rack until clear. Bottle, cork, and allow it to mature.

This really does not vary too much from the first recipe, except for the addition of fruits or flowers.

Spinach Wine]

STRANGE OR AMUSING as it sounds, a very good "green wine," pleasantly and unusually flavored, can be made from spinach, a leaf that has been subject to more jest in our country than any other green leaf. And so, following the popular folk tradition, I add that it will probably give you strong, firm muscles. I will vouch for the warming strength of the wine.

The plant was originally brought from Arabia, and it is interesting to note that it has retained the name, in a measure, the Arabic word for it being *ispinaj*.

For wine making, it is best to use fresh spinach, preferably from your own garden. If you buy it, be certain that you get fresh crisp leaves. Then the wine will have a pleasing green color, for which reason it is called green wine in England, where I drank it first. There I was told that the leaves must be fresh and perfectly dry.

This is a recipe for 1 gallon. If you want more, just multiply the amounts of the ingredients. Don't drink it for at least a year.

You need: *1 ¼ gal. water*
3 to 4 lbs. sugar
2 lemons
2 oranges
4 lbs. spinach
1 lb. raisins
¼ oz. yeast (1 package)

1. Put into an enamel pot the water, sugar (dissolve it well), thin rinds of well-washed lemons and oranges, and finally the spinach, well washed and dried.
2. Bring the water to a boil and let it simmer for about half an hour.
3. Strain through a cloth into a crock, squeezing every drop of juice from the leaves.
4. Cut the raisins the best you can and add them to the crock.
5. Now add the lemon and orange juice, and when the liquid is lukewarm, put in the yeast, dissolved in ½ cup warm water.
6. Cover and let stand in a warm place (65°-70°) until the fermentation has ended. Then strain the wine into glass jars or gallons and let it rest for at least two weeks.
7. Siphon off the clear wine and fine the rest. After fining, I generally pour the wine back into clear gallons and let it stand for another month or more to see if it needs decanting. When absolutely clear, bottle it.

[*Tomato Wine*]

TO BEGIN THIS CHAPTER PROPERLY, I drank a glass of tomato wine that had matured for only eleven months. I also gave a glass to Marion, my ever-laughing, efficient secretary, who is a good wine drinker. We both agreed that it was a good wine—a white-Burgundy-type with a flavor and body uniquely its own, clear and golden yellow. It would go well with fish or chicken and add a zest and savor not to be found in other wines. Next month, when the tomatoes in our garden are red and fiery ripe, I shall make at least two or three dozen bottles. But let me warn you, the taste of the wine is unpleasant when it is first made. It *must* be allowed to mature before you can enjoy it.

I think it will interest you to know that the tomato has gone through a bewildering cycle of ups and downs, "hot and cold," such as no other plant has. It was considered a decorative plant, a fruit, a poison, an aphrodisiac, an ungodly fruit, a cause of cancer, and many other things until it finally became the most popular fruit in North America.

As far as we know, tomatoes were first heard of in Peru. Next we learn of them in Mexico, where they were called *tomatl*. The Spaniards then brought them to Europe and they were grown as a curiosity. The Moors of Spain were particularly fond of the glowing red and glowing yellow, soft-touching fruit, and so, when Italian sailors bought them in Tangier from the Moors, the fruit was baptized *pomo dei Mori*. When the fruit came to France under that name, it became very popular and was translated as *pomme d'amour*, which the English retranslated literally as "love apples." Perhaps because of this mistaken name, it was considered an aphrodisiac.

Then came a change—reputations even of plants are sometimes fickle. The Renaissance painters depicted it as a symbol of poison, and it was considered a dangerous ornamental plant. Thomas Jefferson cultivated the fruit in Virginia. It soon spread through the country, under serious suspicion as a dreadful poison, until Colonel Robert Gibbon ate it publicly on the steps of the court house in Connecticut. From then on, it grew in popularity. And today it is said to be overloaded with vitamins and what not. It has completely overshadowed the muscle-producing spinach.

The Recipes

There are two ways of making tomato wine, one with water and one without. I have made both, and one is as good as the other.

I
Pure tomato wine

You need: *Tomatoes, any amount*
3 to 4 lbs. sugar per pound of fruit
Salt

1. Take any amount of tomatoes you wish. If bought, wash them well. Put them into a crock and mash them thoroughly by hand or with a wooden spoon. Cover and let stand for twenty-four hours.
2. The following day, strain and press out every drop of the juice into an enamel pot. Pour the liquid back into the crock and add a pinch or two of salt.
3. Add 3 to 4 pounds sugar per pound of tomatoes. Cover and put in a warm place (65°-70°) to ferment. It should ferment without

yeast. (It did with my wine.) If it does not, you can always add yeast—¼ oz. (1 package) for each 2 gallons of must. During the fermentation period, a certain amount of froth will come up each day. Remove that.

4. When the wine is still, strain it into glass jars. There let it stand two weeks or more to clear. If it does not, fine it with eggshells or isinglass.

Let me warn you again: Do not drink the wine until it is at least one year old. It is better to wait longer, for it needs maturing in the glass. I tried it raw and it was unpalatable—until a year later.

II

Tomato wine with water

You need: 2 gals. water
6 to 8 lbs. sugar
½ oz. fresh ginger root
16 to 20 lbs. tomatoes
2 tbsps. salt
½ oz. yeast (2 packages)

1. Into the water in an enamel pot, put about 7 pounds sugar and the ginger root, bruised. Bring this to a boil.
2. While the water is heating, thoroughly wash the ripe tomatoes. Cut them into small pieces and put them into a crock. When the water boils, pour it over the tomatoes and add the salt.
3. Dissolve the yeast in ½ cup warm water and pour that into the liquid when it is lukewarm.
4. Cover and put in a warm place (65°-70°) to ferment. Mash the tomatoes each day with your hands or a wooden spoon. Fermentation will take between fourteen and twenty-one days.
5. When the action has ended, strain the wine into glass jars. Let it

stand for two weeks and then siphon off the clear part, if there is any, fining the rest. Rack, if necessary.

6. When the wine is perfectly clear, bottle it. Remember this wine needs at least a year for maturing.

[*Wheat Wine*]

IT IS SAID and it is written by the scholars of Araby that when Adam and Eve were driven from the Garden, they took with them only three things: an ear of wheat, which is the principal of all foods; dates, which are the most important of fruits; and myrtle, which is the finest of scented flowers. Thus, wheat was said to have been cultivated from earliest days, and probably this is a fact. Incidentally, wheat is usually known as "corn" in Europe. The American "corn" is really Indian corn, properly called "maize."

In northern Europe there is a Wheat Mother (or Corn Mother), the spirit of the wheat, and there are many ceremonies and customs associated with her, performed at planting time and at the gathering of the harvest. But in Silesia, in Germany, from which my family came years ago, it is the wolf and the dog who are the embodiment of wheat. The one who binds the last sheaf of grain is called the "wheat wolf" or the "wheat dog." This is also the case in a few other parts of Europe. Often other animals are associated with this custom: goats, cocks, pigs, bears. I have never heard of this in Eastern lands.

Whatever the connotation of the grain, it makes, as all the world knows, a very potent drink, even as a wine, and I can attest to that. Here is how you make it.

You need: 2 *lbs. wheat** (*in grain form*)
2 *lbs. raisins*
2 *lbs. potatoes*
2 *lemons*
2 *oranges*
2 *gals. water*
8 *lbs. sugar*
½ *oz. yeast* (*2 packages*)

1. Put the wheat into a crock, together with the raisins, cut fine; the potatoes, well washed but not peeled, and cut into quarters; and the thin rinds of well-washed lemons and oranges.
2. Heat the water in an enamel pot, dissolving the sugar in it. When it boils, pour it into the crock. Add the juice of the lemons and oranges.
3. When the liquid is lukewarm, dissolve the yeast in ½ cup warm water and add it.
4. Cover well and set in a warm place (65°-70°) to ferment. This will take about twenty-one days.
5. When the fermentation has ended, strain the wine into glass gallon jars and let it stand two weeks or more. There will be a sediment at the bottom. Decant the clear wine and fine the rest.

This wine will have a nondescript, strong vodka-like or marc-like taste. You can flavor it with any wine you like, or sweeten it with syrup, or drink it as it is. But don't do that until it is a year, preferably two years, old. The maturing mellows the wine, quiets the strong flavor, and seems to make it more potent. And a potent wine it is! Try it, and you will find it so.

* You will find wheat in Italian-American, Greek-American, and probably in any grocery that carries Eastern products. Some "health" stores have it as well, and sometimes it is found in the grocery section of large department stores.

HERB WINES

Angelica Wine]

THE HERB WITH THE heavenly illuminated name has not fared too well in its scented realm. One considerably prejudiced herbalist described it as a biennial herb ". . . frequently chosen as a garden subject because it sounds heavenly, but is rather a plebeian-looking plant which masquerades under a pretty name." Yet, having grown it in our garden, I disagree emphatically with the good man. It is not unattractive; rather, it is pleasing to see, with its cone-shaped bluish flowers, tall stems, and generous widespread leaves. It is a decorative plant when standing against a white-and-gray steel-veined stone wall with shorter green herbs in front. It is very easy to grow—all it needs is a shady place.

Its history is honorable and filled with overtones of usefulness.

It was considered an excellent and potent agent against witchcraft and fearsome creatures. So strong a power did it possess that it was even deemed a remedy against the dreaded plagues of the Elizabethan days. It had a proper place among the medicines of the time, and it is still used today for medicine and flavoring. Many beauty lotions have angelica as one of their essentials. It also had an important place as a food. Angelica leaves have a pleasing piny taste when raw; they often went, and still go, into salads. Sometimes they were eaten roasted or boiled. At one time, the stems and leaves were sugared and eaten as candy.

In former days the roots were called "roots of the Holy Ghost." These also were used as a food.

The oil of the angelica plant has an important place in the making of liqueurs such as Benedictine and chartreuse and also in vermouth. Thus it deserves to be called an attractive and useful plant and, for our purposes, a valuable plant, for it makes an unusual and exotic-flavored wine, well worth a place in the cellar.

As far back as 1759 there are recipes for making a wine from angelica leaves. A lady by the name of Elizabeth Cleland described it, and good men spoke of its potency and excellence.

Today it is a wine for those who grow their own herbs, and I believe it can also be made from dried leaves. These can be bought in places that sell dried herbs. The names of some of these companies are listed at the end of the book.

And now I will tell you how to make the wine, even as I have made it.

You need: *2 gals. water*
 2 lemons
 2 oranges
 1 lb. angelica leaves (or 2 lbs., if you want to make it stronger)
 7 lbs. sugar
 ½ oz. yeast (2 packages)

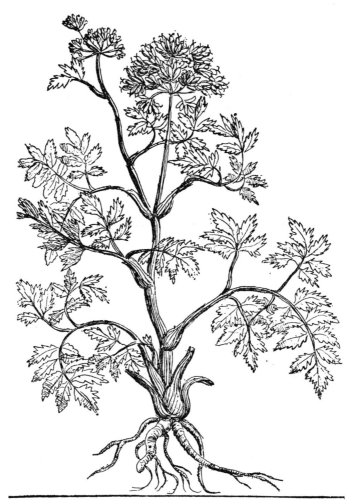

Angelica syluestris maior.

1. Put the water into an enamel vessel.
2. Wash and peel the lemons and oranges very thinly, and add the rinds to the water.
3. Squeeze the juice from the citrus fruit and put that into the liquid.
4. Take the angelica leaves and crush them in your hands. Throw them into the water.
5. Heat the liquid until it boils; boil for about twenty-five minutes.
6. Let it cool a little and then strain it off into a crock. Squeeze the leaves to get all possible juice out, and add the juice to the liquid in the crock.
7. Add the sugar and dissolve it thoroughly.
8. When the liquid is lukewarm, dissolve the yeast in ½ cup warm water, and pour it into the liquid.
9. Cover the crock and set it in a warm place to ferment for about fourteen days. Strain, fine, and rack.

Do not drink the wine for at least one year.

 Balm Wine]

THE HERB, which, according to the vivid words of olden writers, "yields an incomparable wine," has a short, bright history that tells of its pleasant virtues and values. It was often discussed by ancient scriveners, poets, and herbalists in their writings—and with good reason, for it has one uncommon virtue. It has, like the rose geranium, a ravishing fragrance of mint and lemon.

The Arabic herbalist, Ibn Baithar, setting down his observations and knowledge in beautiful Arabian calligraphy, advised the people of his land to dry the complete plant and sew it with a silken thread into a linen bag and wear it under their clothing, for it had the virtue of making one beloved and agreeable. The wearer will have all his wishes fulfilled and will be joyous and happy. Need I add more to explain why I plant it in my garden? Each summer it yields an abundant harvest of leaves, plus a most excellent wine.

Another Arabic author advises us to dip the plant in honey and sugar and smear it on an empty beehive. Bees will then swarm to the hive in great abundance.

Balm, brewed as a tea, was also used frequently in medicines as a cure for asthma, fever, and headache.

John Parkinson, in his well-known *Theatricum Botanicvm; The Theater of Plants,* advises its use as a "tansy" with eggs, sugar, and rose water to increase the flow of a nursing mother's milk.

It was also highly recommended in foods and to flavor cold drinks, and it still is today. Salads, teas, and sauces for meats gain from its delicate presence.

Balm was popular in "pomanders"—those compounds of many scented herbs, popular with our great-grandmothers. The herbs were pounded together and then arranged in artful shapes and designs. Our present-day oranges, lemons, or apples studded with cloves are reminders of old pomanders.

The Recipes

I

First let me acquaint you with a recipe suggested by Elizabeth Moxon in her *English Housewifry.* I have modified it slightly so that it can be used today.

You need: *4 qts. balm leaves, slightly packed down*
2 gals. water
4 lbs. sugar
whites of 4 or 5 eggs (optional)
½ oz. yeast (2 packages)

1. Crush the balm leaves and put them into a 4-gallon crock.
2. Put the water into an enamel pot, bring it to a boil, and pour it over the crushed leaves.
3. Let it stand for twenty-four hours, then strain it off into an enamel crock, squeezing all the juice from the leaves.
4. Add the sugar to the crock and stir with a wooden spoon until it is well dissolved.
5. (Here Mrs. Moxon suggests that one put in the whites of four or five eggs and "whisk it very well before it be overhot, when the skin begins to rise, take it off, and keep skimming it all the while it is boiling. Let it boil three quarters of an hour, and then put it in the tub.") I eliminated this step and found the wine no worse for it.
6. Heat about half of the liquid separately, then pour it back into the crock.
7. Dissolve the yeast in ½ cup lukewarm water and put it into the crock.
8. Cover the crock and set it in a warmish (65°-70°) place. It will take fourteen to twenty-one days to ferment. When the visible fermentation has stopped, strain the wine, clear it, fine and rack if necessary, and then bottle.

II

Sir Hugh Plat, in 1602, told us how he made a balm wine, as set down in his charmingly written book: *Delightes for Ladies*. He spoke of "distilling" the wine. I am not interested in "distilling";

BALM

CARAWAY

but the recipe gave me an idea, and so, prompted by curiosity and interest in wine adventure, here is what I did.

1. I took half a gallon of domestic Rhine wine and put it into an enamel pot.
2. Then I put into the wine ½ pound macerated balm leaves.
3. I started the burner under the liquid and let it simmer for ten minutes.
4. Next, I put the wine back into the gallon bottle, closed it, and put it away. Five weeks later I had a delightful dessert wine with a lemony-mint flavor.

III

Here is a simple, tried recipe.

You need: *4 lbs. sugar*
2 gals. water
2 lemons
3 qts. balm leaves (not pressed down tightly)
1 lb. raisins
½ oz. yeast (2 packages)

1. Dissolve the sugar thoroughly in the water, using an enamel vessel.
2. Wash the lemons. Cut the rinds very thinly and put them into the water. Squeeze the juice and add it. Then put it all on the fire and bring to a boil.
3. Crush the balm leaves in your hands and put them into your crock. Pour the boiling-hot liquid (not strained) onto the leaves.
4. Add the raisins, cut as small as you can. Let the liquid cool.
5. Dissolve the yeast in ½ cup lukewarm water and pour it into the lukewarm liquid. Cover; let stand in a warm place (65°-70°).
6. After four days, strain the fermenting liquid off the leaves and rind, and let it ferment until all the violent fermentation has ended.

7. Now strain the wine into glass jars, let it stand to clear, or fine if necessary, and bottle. Let it rest at least a year.

As with most wines you make, you may introduce a lump of sugar into the bottle to produce a slightly effervescing, *pétillant* wine, which brings a piquant taste and an unusual bouquet.

You can try to make balm wine with dried leaves, bought commercially from herb dealers, both in this country and in England.

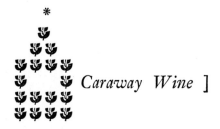 Caraway Wine]

PERHAPS IT IS NOT appropriate to begin the description of an herb wine by speaking of the herb's edibility, yet I cannot help but say that it is a great pity that the caraway root is not as popular as the seed, so much in favor for flavoring. For the root is sweet to the taste and perhaps more pleasantly flavored than parsnips, and can be used in the same manner. Even the leaves, which have an agreeable odor and a tangy taste, are more pleasing than the seeds.

In colonial days, the seeds were often used as chewing gum is today—to sweeten the breath and to keep the mandibles busy.

But the use of caraway antedates our colonial grandmothers. The seeds have been found in the debris of the ancient lake dwellings of Switzerland.

Ashurbanipal, ruler of many lands around 650 B.C., famous for a magnificent library of tablets, which he collected wheresoever he could lay his hands on them, had among them the famous Hammurabi Code from Nineveh, chiseled out around 2100 B.C. In that code mention is made of caraway seeds. The good and wise people of those days also took a little caraway wine, which they made as a stimulant.

Among the Egyptians and the Israelites, the seed was most popular as medicine in the form of tea. It was excellent for stomach difficulties.

The Ebers Papyrus, found among the bones of a mummy in the Theban necropolis, speaks of caraway.

Quite a few Roman writers, among them Apicius and Pliny, also mention caraway. The Romans mixed the roots with oil and made a bread of it. The great Dioscorides prescribed it for pale-faced girls who desired pink cheeks.

Charlemagne, Ibn Baithar, and many others speak of these seeds. Caraway was an important herb, according to the great seventeenth-century Dr. Stephens. He prescribed a "sovereign water" containing caraway as an incomparable cure-all to keep one alive for endless years, for ". . . it comforteth the spirits and vital parts and helpeth all inward diseases that cometh of cold, and whoever useth this water moderately will make him seeme young in old age. This water will better if it be set in the sunne all the summer."

Caraway seeds were popular and were used in many other "waters," cordials, etc.

Today Scotsmen make a fine caraway whisky, which is used without end in cooking and baking. I'll tell you about it in the chapters on brandies and unusual drinks.

I like caraway wine (the taste of which is probably familiar to you as kümmel) because it is pleasant-tasting and it brings back the sunny days of my young years when I drank it every Friday. I don't know whether my mother bought it or made it, though I vaguely remember seeing her put sugar syrup and caraway seeds into a white liquid.

The Recipe for the Good Wine

You need: 2¼ gals. water
5 to 7 lbs. sugar
1 lb. barley (any kind)
½ oz. yeast (2 packages)
1 slice toast
2 oz. caraway seeds, soaked in ½ cup brandy (or vodka, gin, barley wine, potato wine)

1. Dissolve the sugar thoroughly in the water. Then add the barley. Boil this for a full thirty minutes, skimming off whatever impurities come to the top.
2. Strain the liquid into a crock and let it cool.
3. Dissolve the yeast in ½ cup lukewarm water. Cover the toast with it and place both in the crock, pouring in whatever yeast liquid remains. Let the must stand for twenty-four hours, stirring it very gently once or twice.
4. The next day put the caraway seeds into a little brandy or one of the other spirits mentioned above. Soak the seeds for about four hours; then add them to the crock.
5. Let the must remain quiet until fermentation stops, which will take from ten to twenty-one days.
6. When active fermentation has ended, strain the wine into glass jars or gallons and let it stand to clear. If it does not, fine and decant. Then bottle and cork.

In a year, you will have a pleasant and full-bodied wine.

 Chickweed Wine]

I LEARNED OF THIS WINE while I was at work (together with C. H. Tillhagen of Sweden) on a book of gypsy folktales.*

While motoring through France, I came upon a whole gypsy caravan at rest. Their wagons were strikingly colored, and the women and children wore multicolored clothing. The men were weaving baskets. I stopped, gave the youngsters candy and coppers, took photographs, and entered into conversation with the older folks. From them I learned that they made wine from *mouron* (chickweed) that the youngsters gathered. It was also cooked as a vegetable. In parts of England, it was boiled and eaten to clear up acne and to give nice pink cheeks.

If you live in the country and are cursed with chickweed in your flower or vegetable garden, just don't weed for a week and it will spread along the ground like fire in dry grass, giving you all that you want.

If you live in the city and are an adventurous viniculturist and would like to make this unusual wine, go out to the country to any farm and tell the owner you would like to pick chickweed in his garden. There is always chickweed in a garden. He will think you have gone mad, but I am sure he'll grant your request. And so you will gather the weed and proceed as follows.

* *The Gypsies' Fiddle*. Vanguard Press, New York.

You need: *2 lbs. chickweed (2 or 3 full quarts, loosely packed)*
2 gals. water
2 lemons
2 oranges
6 lbs. sugar
1 slice toast
½ oz. yeast (2 packages)

1. Put the chickweed (without the roots) into the water in an enamel pot. Add the thin rinds of the lemons and oranges, and boil together for fifteen to twenty minutes. Strain the liquid into a crock.
2. Dissolve the sugar thoroughly. Add the juice from the oranges and lemons.
3. Prepare a slice of toast. Cover it with the yeast, dissolved in ½ cup warm water. Add both to the crock.
4. Let it ferment for fourteen to twenty-one days. Then clear, decant, and bottle.

If you are not pleased with the flavor of the wine and want to change it, you can do so as suggested under the heading "Taste and Flavor," on page 57.

[Clove Wine]

I FIRST MADE CLOVE WINE as a pleasant and perfect accompaniment to dark brown honey cake, and my friends and I found much pleasure in drinking it. Then I decided to include it in the book and thus give other wine adventurers an opportunity to make it.

The magnificent part these pungent, spiced kernels have played in the march of history, in kindling wines, in sumptuous perfumes, and in ceremonial oils of days gone by, entitles them to a broader life than just as a food flavoring. So it is fit and proper to bring them into the rich, sovereign realm of viniculture.

The Persians ("Persians" is a much more beautiful sounding word than "Iranians"; I am sure no poet changed the name) used cloves in recipes to rekindle lost affections. It was particularly popular among married women to bring back errant husbands. This is so charming a concept and formula that I cannot help putting it down for the benefit of some modern ladies who may like to use it:

Place cloves, cardamon, and cinnamon in a jar and read over it several times, backwards, the Yā-Sīn sura of the Koran. Then fill the jar with rose water. Put the husband's shirt in the liquid, and also a piece of parchment that has his name on it and the names of the four angels [probably the archangels]. Put it all over a fire, and as the mixture boils, the errant husband will return.

Of course, cloves were used in love potions, and there are endless recipes for them. They are also part of Dr. Stephens' famous recipe as a cure-all for the ills of the world and for rejuvenation.

Strong, spicy odors always have been considered potent to prevent the harms of witchcraft; and cloves are naturally favorites in this field.

Among some primitive peoples, the clove tree was of prime importance. In the Moluccas, the islanders treat the blossoming clove tree with the same consideration with which they treat their pregnant women. All noise is prohibited near them. Neither bright lights nor garish fires may be lit near them in the dark of the scented night, lest they be unduly disturbed. Alarming incidents near them must be avoided, so as not to upset the maturing of the fruit. The blossoming time is such an important event that men, passing, must remove the covering from their heads out of sheer respect for the coming fruit.

The clove, as an ornament and for its lovely strong scent, was a favorite in making "spice oranges" and "spice apples." Such fruit, stuck all over with the pleasant brown-black kernels, is common even today.

The clove is also used to this day in certain religious ceremonies of the Jews, one phase of which is to smell the scent.

Wine spiced with cloves has been very popular ever since the Middle Ages. In England, gin-and-cloves was popular during the eighteenth century. Alas! Those rich, Oriental-scented drinks have been driven away by brimstone liquor that burns the taste buds and blears the brain cells.

Here is how you can make your own pleasant Eastern folk drink. Try it for its rich history and for its delightful spicy taste and bouquet.

The Recipe

You need: *2¼ gals. water*
3 oranges
3 lemons
1 oz. fresh ginger root
2 oz. cloves
5 to 6 lbs. sugar
¾ oz. yeast (3 packages)

1. Put the water into an enamel pot. Wash and peel the oranges and lemons, and put in the rinds. Then add the ginger and cloves. Bring to a boil slowly and let the liquid simmer for at least an hour; a little longer will do no harm.
2. Strain into a crock and add the sugar, dissolving it well. Slice the oranges and squeeze the juice from the lemons and add both.
3. When the liquid is lukewarm, add the yeast, dissolved in ½ cup warm water. Cover and put in a warmish place (65°-70°) to ferment. This should take about fourteen days, perhaps more.

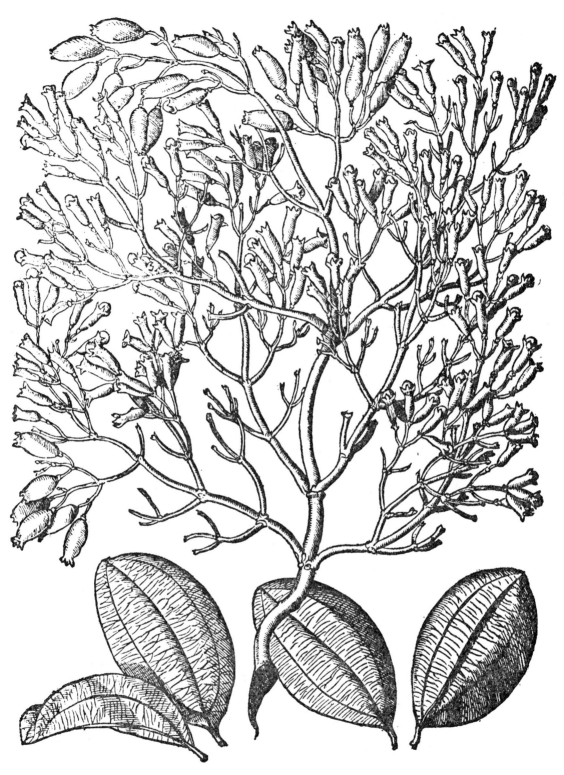

CLOVE

4. Clear, fine, and bottle.

This wine can be drunk within a year.

[Dill Wine]

"TREFOIL, verain, John's wort, dill,
 Hinder witches at their will."

So runs an old rhyme; and perhaps one of the reasons I grow the lovely, tall, slender, swaying stems with greenish-yellow flowers looking in all directions is to keep the witches away. Of witches, there are none in my garden; but bugs, eating and destroying, there are aplenty. Would there were some green leaves or colored flowers to keep those away!

The pleasant aroma and taste of dill bring to me memories of my earliest years, when my mother and the mothers of my play companions dried it in the open and used it in pickling cucumbers. And children and grownups have smelled the same pungent scent and tasted the same pleasant flavor for thousands of years before me. Dill was used in Biblical days and was a favorite in the Far East. To this day, you smell it in open market places, wheresoever there are open market places left, whether in the Far East or in the Mediterranean countries or in other parts of the world where it grows.

The Romans liked it so much, they put it into the scented flower crowns they wore around their heads.

It was Pythagoras who proclaimed that holding a stem of dill in the left hand would prevent fits. The great King Charlemagne favored it in food, even as I do.

Down the gray, medieval years, dill was used in medicines, love potions, and brews for frightening spirits. It was very popular, especially for this last purpose. All flowers with yellow corollas, which were said to be directly connected with gold and the sun, were forces with which to combat witches. Farmers were advised to rub dill and salt on their newborn calves to keep them safe. And the beauticians around the 1800's advised ladies and gentlemen who wanted slim waistlines to boil dill in a broth and drink it as often as possible.

So through the ages, to this very day, dill has been popular among those who like pleasant scents and flavored foods. It is so popular that one type of spiced cucumber is called "dill pickle" because of the use of dill, which gives it a tantalizing aroma. And I assure you that it also makes an aromatic and very flavorsome wine, which will bring you a rare gustatory experience.

The Recipe

You need: *1 oz. dill seed*
1 cup brandy
2 gals. water
1 lb. rice or barley
6 to 8 lbs. sugar
½ oz. yeast (2 packages)

1. Take the dill, either purchased or grown in your own garden (and dried in an airy, shaded place—not in the sunlight); put it into the brandy and let it soak for about twelve hours, preferably overnight.
2. Put the water and the rice or barley in an enamel pot and boil for fifteen minutes, taking off any scum that rises.
3. Strain the liquid into a crock. You can use the cereal as food.

4. Dissolve the sugar in it. (If you like a sweet wine, use 7 or 8 pounds.)
5. Add the dill and the brandy.
6. Dissolve the yeast in ½ cup warm water and add it.
7. Let the must ferment from ten to twenty-one days; then strain, clear, and bottle.

Don't drink the wine for at least a year.

 Ginger Wine]

I COULD NOT REFRAIN from including a ginger wine for you to make, since ginger is recommended so often in wines and has been used so much throughout the greater part of the world in cooking and flavoring. In China, and probably in Japan, it is a pavilioned household word. And it gives a rich, heavy wine, a pleasure after dinner and a splendid accompaniment to the poems of Li Po or the travels of Marco Polo—or any other book you like.

In earlier days, ginger served primarily as a drug—both as a healing drug and as a drug for the scarlet malady of love. There were innumerable love potions in tablet and powdered form, containing unusual and rare additional ingredients: dried crocodiles, birds' tongues, and others of the same ilk. Should you like to make one, here is a recipe:

"Take cinnamon, ginger, pepper, cress seed, rocket seed, mustard seed, of each half a drachm; bird's tongue, onion seed, crocodile, of each one scruple; white sugar dissolved in rose water, four ounces. Make into tablets."

Ginger was also an ingredient in "water of thyme"—good for the passions of the heart.

Pythagoras used ginger in his antidote for poisoning. In later years, it was part of Dr. Stephens' prescription cure for all the ills of the world and for prolonging life. One of the ever-recurring tragedies of the Middle Ages was the plague, and ginger was one of the endless cures given: ". . . a spoonful of it with wine and herbs, and you shall be safe for twenty-four days. Nine times taking of it is sufficient for a whole year, by the Grace of God." This prescription was sent by Henry VIII to the Lord Mayor of London.

But as time went on and the plagues went away, ginger served other purposes—as a spice in foods, and such it has remained to this day. Now you can also use it as a wine, and this is how.

The Recipe

You need:
- *9 lbs. sugar*
- *3 gals. water*
- *1 lb. raisins*
- *8 oz. fresh ginger root*
- *4 lemons*
- *4 oranges*
- *¾ oz. yeast (3 packages)*

1. Dissolve the sugar in the water. Add the raisins and ginger, both cut into pieces.
2. Bring slowly to a boil and simmer for a full hour.
3. Strain the liquid into a crock.
4. Wash the lemons and oranges well; peel the rinds very thinly and put rinds and fruit juice into the crock.
5. Dissolve the yeast in ½ cup warm water and add it.

6. Cover, put in a warm place (65°-70°), and let it ferment from ten to twenty-one days.
7. Strain into glass jars or gallons. Wait two weeks to see if it clears. If it does not, fine and rack.
8. When clear, bottle and cork.

In a year you will have a fine dessert wine.

 May Wine]

THIS IS A VERY POPULAR SPRING WINE, and justly so, because of its pleasant and delicate bouquet. Its sweet perfume comes from the herb used in it: sweet woodruff.

There are two kinds of May wine, the commercial type and the homemade type.

I
The Commercial Wine

Get a bottle (or any number of bottles) of white wine on the sweet side, a sauterne or even a Châteaux d'Yquem. Put into it a few sprigs of fresh woodruff, or use the dried leaves, which can be bought. After a week, strain and drink.

I use woodruff grown in my own garden, and I let the herb stay in the wine until the bottle is finished.

II
Homemade Woodruff Wine

You need: 2 gals. water
           2 handfuls woodruff, fresh or dried
           2 lemons
           5 to 6 lbs. sugar
           ½ oz. yeast (2 packages)

1. Put into the water half the woodruff and the lemons, washed and sliced. Boil for twenty minutes.
2. Strain into a crock. Add the sugar and dissolve it. Then throw in the remaining fresh or dried woodruff sprigs.
3. Dissolve the yeast in ½ cup lukewarm water and pour it into the liquid when it is lukewarm.
4. Cover and let stand in a warmish place (65°-70°) until the fermentation has ended. This will take about two weeks.
5. Strain the wine into gallon jars and let it stand for two weeks. It should be clear by then, but if it is not, fine it.

You can regulate the perfume and the sweetness of the wine to suit your own tastes. If you like a stronger scent of the herb, put a sprig or a few dried leaves into each bottle.

The wine can be drunk eight to twelve months later.

WOODRUFF

 Mint Wine]

BEFORE I BEGAN writing these lines, I drank a libation of two full wine glasses of fragrant two-year-old mint wine, and "it was good," as it says in the Bible. Now I feel in fine fettle to write about it.

Mint, the symbol of wisdom, has a history and lore as rich as its flavor and scent, so greatly prized in flavoring and cooking. As a matter of fact, mint played a much greater part in the daily lives of the Greeks and Romans than it does today. Even before them, the Israelites strewed mint leaves and stems on the floors of the temple and of their homes to give a cooling, pleasant fragrance as they walked over them—a custom that was followed later in churches.

Still later, the fanciful writers, following the ancient custom, said that "Mint should be smelled as being refreshing for the head and memory," and so again the ancient custom was revived and mint was strewn "in chambers and places of recreation, pleasure and repose, and when feasts and banquets are to be made."

With the customary poetic cadence of their literary (and daily) life, the Greeks built a beautiful tale around Minthe.

Pluto fell in love with Minthe, a daughter of Cocytus. When Mrs. Pluto, Proserpine, learned of this, she was bitten by furious jealousy, as is the unreasonable habit of women, and turned the little nymph into a plant. But she could not destroy the fragrance and the freshness of the young maid, and so Minthe has remained mint to this day.

The Romans were great lovers of the herb. They used it in foods, as perfume, and for many other purposes. Roman women mixed it with honey to form a candy paste, which they ate to obliterate the scent of wine from their breath—wine being forbidden to them, under penalty of death.

It would be most interesting to learn what anger prompted which Roman senator to have this ungentle law passed! I am sure he perfumed his bath with mint leaves, as was the custom of the times.

Pliny has these words for mint: "The very smell of itself refreshes and recovers the spirit as the taste inflames our appetite for meat, which is the reason why it is so much used in our sauces, wherein we are accustomed to dip our meat."

Marcus Gavius Apicius, a great epicure around 14-37 A.D., was an ardent mint fan. He speaks of it for pages and pages in his famous cookbook. Dioscorides praised it for its value in curing stomachache, and it is still used for that purpose.

Medieval imaginative herbalists and scholars, in the eddy of their creative turmoil, put their own trellises around the scented plant. Said they: "The root, if chewed, makes a most gentle person fierce and quarrelsome, even as catnip makes cats dance, fighting and full of love."

In France, spearmint is dedicated to the Virgin. It is called "menthe de Notre Dame" (Our Lady's mint). And in Italy, too, it is called "erba de Santa Maria." And so, with classic and ecclesiastic sanction, to the wine!

The Recipe

A word of warning to those who wish to grow their own mint. Once you plant it, it is difficult to curb—unless you put tin (from old tin cans) or some other metal about six to eight inches deep around the mint bed. The roots are very shallow and will not go beyond the metal wall.

Whether you use purchased or home-grown mint, it makes a fine, cooling wine, perfect for a hot day. I make three gallons at a time, and you should too. Your friends will beg for it, and you will like it. If you want to make less, decrease the amounts of the ingredients proportionately.

You need: *6 to 9 lbs. sugar*
3 gals. water
3 qts. young mint leaves, packed down rather firmly
1 slice toast
½ oz. yeast (2 packages)

1. Dissolve 6 to 9 pounds sugar (depending on how sweet you want your wine) in the water in an enamel pot. Heat to boiling.
2. Put fresh mint leaves into a crock, crushing them in your hands as you do so.
3. Pour the heated syrup over the crushed leaves.
4. Prepare a slice of toast. Cover it with the yeast, dissolved in ½ cup warm water. When the liquid is lukewarm, add the toast and remaining yeast to the crock.
5. Cover well, put in a warm place (65°-70°), and let ferment for about two weeks or a little longer.
6. When the fermentation has ceased, strain the wine into glass jars or gallons. Let it stand for two or more weeks to clarify. If it does not, fine with eggshells or isinglass. Let it stand at least another week or two, then decant if necessary and bottle.

MINT

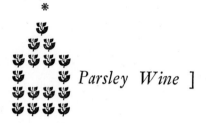

[*Parsley Wine*]

LET ME INTRODUCE YOU to a wine with a rare, fantastic taste that becomes more pleasurable the more you drink. Not only is it rare, excellent, and stimulating, but it is famous through herbal medicine as a specific against arthritic pains—which seem to afflict every other person I know. But it is the flavor and the bouquet of this wine that fascinate me. It was completely new, delicate, haunting, provocative, and when prepared with other herbs—of which I will tell you later—became a creation of truly Oriental wizardry with a bouquet and taste out of a great poet's imagination. One might call it a mysterious iridescence of flavor and taste.

I just drank a glass of it, and it runs through the body, creating a glittering warmth and taste, on the fine edge of irritation, because of its completely exciting newness, comparable to no other taste.

Parsley was an important plant to the Greeks and the Romans. Garlands of parsley were the prizes often awarded by the Greeks to those who won athletic contests, given in honor of a friend who had just died. The reason for that presumably was that when Archemorus, "forerunner of death," one of their fabled heroes, died, a parsley plant grew out of his blood.

For the same reason, it was usually one of the important decorations of tombs. From this arose the quaint expression among the ancients: "in need of parsley," said of a person who was incurably ill.

I wonder if our modern habit of crowning with parsley a joint of meat or other foods is not in a measure a descendant of the ancient Greek custom of crowning with parsley.

To turn to a lighter vein, Plutarch, the great historian, relates that once a cunning enemy sent a few mules all covered with parsley into an army of Greeks who had come to conquer—and the Greeks turned and fled in great fear!

The name we use comes from the Greek, meaning a plant that grows on a rock. From this came the Latin, *petroselinum*, which underwent many changes in French and German and finally became our word parsley. It was the Romans who introduced parsley into England, whence it spread everywhere.

Like other herbs, parsley had a high place in medieval medicine. It was a sovereign cure for baldness: "Powder your pate with powdered parsley seed three nights every year and the hair will never fall off." It was used for many other remedies as well. Then it was the good fortune of gastronomers to discover its delicate flavor in foods, and of naturopaths to believe that it was excellent for rheumatism, and so the curly little herb became an omen of good flavor and good health—not of death, as it was in the Greek and Roman days. I will add my mite by showing you how it can be also a cause of good cheer —in a good and rare wine.

The Recipes

I

You need: *2 lbs. parsley (2 glass quarts filled but not pressed down)*
          2 gals. water
          2 lemons
          2 oranges
          4 lbs. sugar
          1 slice toast
          ½ oz. yeast (2 packages)

1. Put fresh parsley leaves into the water in an enamel pot.
2. Pare thinly the rinds of the lemons and oranges and put them in. If oranges are out of season, you can double the number of lemons.
3. Bring the liquid slowly to a boil and let it simmer for a full thirty minutes.
4. Strain into a crock, squeezing the parsley through a cloth.
5. Add the juice of the lemons and oranges and the sugar. Dissolve the sugar well.
6. Prepare a slice of toast. Cover with the yeast, dissolved in ½ cup warm water, and add both to the crock. Cover; set the crock in a warm place (65°-70°) to ferment.
7. When the fermentation has ceased, strain the wine into glass jars and let it stand to clear. If it does not clear, fine with eggshells or isinglass, siphon off, and bottle.

II

You need: 2 lbs. *parsley*
2 gals. *water*
4 *lemons*
4 *oranges*
6 lbs. *sugar*
1 *slice toast*
½ oz. *yeast (2 packages)*

1. Put the parsley into a crock.
2. Boil the water and pour it over the herbs. Cover and let stand for a full twenty-four hours.
3. Strain the liquor and return it to the crock, squeezing the parsley through a cloth.
4. Put the thinly pared rinds and the juice of the lemons and oranges into the liquid.
5. Add the sugar; dissolve it well.
6. Heat at least half the liquid and pour it back, so that the whole becomes lukewarm.

7. Proceed as in recipe No. 1.
8. When the fermentation stops, strain the wine into glass jars and let it rest. If it does not clear, fine, decant, and bottle.

The wine gains in bouquet and flavor with time. It is supposed to have great curative value for rheumatism; and if you have no rheumatism, drink it as a preventive.

With either of these wines, as I said at the beginning, you will experience a new, delicate flavor that no other wine has—a flavor difficult to describe. Perhaps you might speak of it as a finely wrought, slightly biting, bitter-nut Far Eastern tang, cunningly blended with the warmth of the wine. It grows pleasurably with repetition. It is well worth experiencing.

III

Perfumed Parsley Wine (parsley–pineapple-sage– mint wine)

Now I will bring to you a drink of wine, rare and exotic, quickening with ancient magic.

I told you that my good lady grows a fine herb garden, and she, too, is adventurous. By sheer accident, she grew a few pineapple-sage plants. One day, when the leaves were in rich summer bloom and I browsed around searching for new, glittering perfumes I find in leaves, I accidentally came upon the pineapple-sage with its long, full leaves. Crushing one, I smelled a perfume lovely, sweet, and rare, to make "the winds love-sick."

I was then in the midst of making parsley wine, and, by a flashing inspiration from Bacchus, I decided to make a wine using the pineapple-sage leaves and some mint leaves, together with the parsley.

So I gathered 2 pounds parsley, and then I picked about 2 handfuls of pineapple-sage leaves and a handful of mint leaves, and then proceeded exactly as I did in the second recipe. The result was a crowned wine, a flavorsome creation of scented herbs that gave a tingling joy to body and mind. A flavor that can only be told in surging litanies of medieval music. Try it, drink it, and see if you don't think and feel as I do about it.

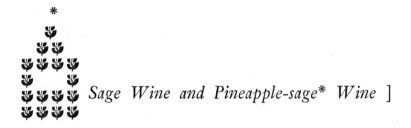 Sage Wine and Pineapple-sage* Wine]

Cur morietur homo cui salvia crescit in horto?
(How can a man die who grows sage in his garden?)

Thus the proverb from the famous school at Salerno, but there are other reasons for making a wine from this herb. It will create a drink with an astonishing and fascinating bouquet.

Sage has been highly esteemed for thousands of years. Pliny and Theophrastus tell of it. Its very name, Salvia (from *salveo*, meaning "I save," because of the curative properties of garden sage), also relates to the way in which sage saved the Blessed Virgin. When she began her flight to Egypt, she hid from Herod's soldiers among the sage plants and was not seen by her pursuers. She blessed the plant, and ever since then the leaves have been fragrant.

The Greeks used sage for flavoring foods and for its medicinal value. It was a favorite flavor for their cheese. (We use it for the same purpose to this day.) The flavor is very strong, and the herb should be used with discretion, in foods as well as in wine.

As I mentioned before, sage has a most unusual fragrance and

* *Salvia rutilans.*

flavor, an almost indescribable mixture of pungent thyme and slightly resinous pinyness.

I once heard or read a "sage" tale, I don't remember just where, of a Greek nymph who lived alone and happy in a hollow oak beside a pool where lots of sage grew. One day, a king passed by, was attracted by the sage, saw the lovely nymph, and fell in love with her. She loved him, too, even though she knew that to love a mortal meant her death. When he took her in his arms, she fainted, and when he rushed to the pool for water, she slowly faded away.

Sage was one of the famous remedies during the Middle Ages. Folks believed that it prolonged life, brightened the spirits, eased sorrow, kept toads away, averted chills, and enabled girls to see their future husbands. It was also held that it grew best for the wealthy and where women were the bullies in the home. Turner, the great English herbalist of olden days, tells of its many healing properties. . . . And so to our wine.

The Recipes

I

You need: *1 to 4 qts. sage leaves*
            2 lbs. raisins
            2 gals. water
            6 to 8 lbs. sugar
            ½ oz. yeast (2 packages)

1. Pick 1 to 4 quarts sage leaves (depending on how strong you want the scent and flavor), and put them into a crock.
2. Cut the raisins finely and add them.
3. Boil the water and sugar together. Pour the boiling syrup over the sage and raisins. Stir well, again and again, with a wooden spoon, macerating the sage and squeezing it against the sides of the crock.
4. Dissolve the yeast in ½ cup warm water and pour it into the

liquid when it is lukewarm. Cover and let stand in a warm place (65°-70°) to ferment. This should be finished in ten to fourteen days.
5. Strain the wine into glass jars and let it stand for two weeks. Then siphon off the clear wine and fine the cloudy.

This wine should not be drunk until it is at least one year old.

II

I have spoken, under parsley wine, of perfumed parsley wine, where I combined parsley, pineapple-sage, and mint. This was perhaps the most unusually flavored and scented beverage I have ever made. A wine using pineapple-sage only will give the same results, and I recommend it to those who revel in exotic flavored drinks. This one is difficult to describe. It can only be likened to a libation reflecting the medieval star-charmed tapestried days of Florence.

The recipe is exactly the same as that used in making sage wine. Pray, do make it, to learn its magic web of flavor and savor.

BERRY WINES

Blackberry Wine]

THE COMMON HIGH-BUSH blackberry has a most honorable and distinguished history: celestial, secular, and Luciferal. It has the ever-crowning glory of being designated as the Biblical burning bush, golden-flamed but never consumed, in which the Lord appeared to Moses. On the other hand, there is the tale of the good Saint Michael, who defeated Lucifer in a great battle in a field of blackberry bushes; and in revenge, the Evil One cursed the innocent, good Rosaceae family with black curses. He stomped his black-hoofed foot all around the patch so that no berry grew there again. Then, to crown his ignominious deed, he covered the bushes with his brimstone cloak and so scorched them that the poor little green leaves turned to a burning red.

All this took place on Saint Simon's day, the 28th of October, which shows what kind of coward Lucifer was. When he couldn't put up a good fight against the great, gleaming archangel with widespread wings, he took it out on a defenseless little bush.

From that historic incident grew a garland of folktales and customs in England and in the United States of America.

In many parts of England, the 10th of October is the last day on which blackberries should be gathered. Note how the date has shifted, but it is still Saint Michaelmas Day. After that date, Lucifer returns to the berry bushes, stomps on them, spits on them, dances on them, and performs all kinds of devilish antics to make the berries unfit for eating, for pies, or for wine.

In Ireland, picking blackberries after that day invites dire punishment. If you do eat them then, in the face of gospel lore, you may die or one dear to you may get into sore straits. So beware of picking and eating blackberries after the 10th of October in England, Ireland, Scotland, or Wales.

Cornwall is fortunate not to be affected by the curse. There, scalds and burns are cured with wetted blackberry leaves.

The medicinal value of this fruit is famous far and wide, particularly in the realm of dentistry. Say the great Welsh physicians of Myddvai:

"To Extract Teeth by Means of a Powder

"Take the roots of nightshade (*solanum nigrum*) and pound them well in goat's milk, then add the blackberries separately pounded thereto, incorporate the whole into the pulp and macerate in vinegar for 13 days; renew the vinegar 3 days, then powder the residue and add vinegar thereto for 3 times more. When this has cleared, decant the vinegar, and dry the sediment in the sun or near the fire, in a like heat. Let the powder be put in a tooth, if there be a cavity therein, and it will extract it without pain and without delay."

Thus teeth can be extracted without forceps, just by the force of this fine powder!

Here is another gem for the dental profession, from *A Curious*

Herbal by Elizabeth Blackwell, published in 1739. If you want loose teeth to become tight again, eat the raw fruit and leaves of blackberries. This is also recommended as a cure for rheumatism.

In some parts of England the berries are called "bumble kites" —the Lord knows why!

These are just a few gray pearls in the oceanful of medicinal and other lore of blackberries.

In our own land, too, the *Rubus allegheniensis*, which is the Latin name of our berry bush, has an important place. The Indians dried the berries, pounded them, and ate them with meat and corn. They ate them raw, of course, and made them into sauces and even jams, as we do. They used them with every kind of food and flavorings. The Iroquois Indians used them as a ceremonial food after soaking them in water and honey.

A charming Indian tale of the blackberry winds around Lake Huron. Once the Frost Spirit, Hatho, came prematurely to that region, destroying the crops. But the Indians drove him off with the boiling juice of blackberries. Since then, before coming to Lake Huron, Hatho looks carefully to see if any blackberries are still on the vines.

The first immigrant settlers were quick to learn the virtues of the berry, and used it for many purposes, among them, to make good wine. This proved a sovereign remedy against winter colds and was used by one and all for that medicinal purpose. Since colds were prevalent among the settlers in those days, the drinking of this good wine was very popular in the new land. If you want to live in the fine old American traditional manner, I recommend that you begin drinking good blackberry wine; and since it is said to be a good preventive of colds, you will want to drink it all the time!

The Recipes

Blackberry wine is one of the three favorite folk wines in America. (The other two are elderberry and dandelion wines.)

I found recipes in old books, and received many for blackberry wines and cordials from all over the land. This popularity comes from the excellent taste of the berry and the abundance of its growth. It is found wild in almost every section of our country.

Like most fruit wines, this one, too, can be made either dry or sweet, depending upon the amount of sugar you put in. (Remember, the yeast needs only a certain amount of sugar for fermentation; the rest sweetens the wine.) A dry wine is good with red meat or a crisply browned duck or goose. When sweet, it is a pleasure to sip while reading a book of good poetry or the essays of Walter Pater. Of course, it is also a pleasure to drink while reading *your* favorite author.

Here is a most excellent recipe sent me by that connoisseur of fine books, Jene Wagner of Brownwood, Texas. But first I should mention two very popular American legal pastimes: (1) making laws —each year we make millions; (2) breaking them. This dates back to our earliest days.

Well over half a century ago, a good citizen of Denison, Texas, complained bitterly in the Texas Almanac about the cussed prohibition laws of the Lone Star State that prevented law-abiding citizens from making good wines. But a short time before that, the Ladies' Association of the First Presbyterian Church of Houston, Texas, published a good cookbook (1883) in which there is, among other wine recipes, one telling how good blackberry wine and cordial can be made! That is the one Jene sent me. It is simple and practical.

I

To every gallon of berries you gather, add a quart of boiling water. Let it stand for twenty-four hours, stirring occasionally.

Strain off the liquid into a cask, and to every gallon add 2 pounds sugar. Cork tightly and let it stand until the coming October, and you will have a wine for use without further straining or boiling. It may be improved and perhaps will keep better if you add a little pure French brandy.

BLACKBERRY

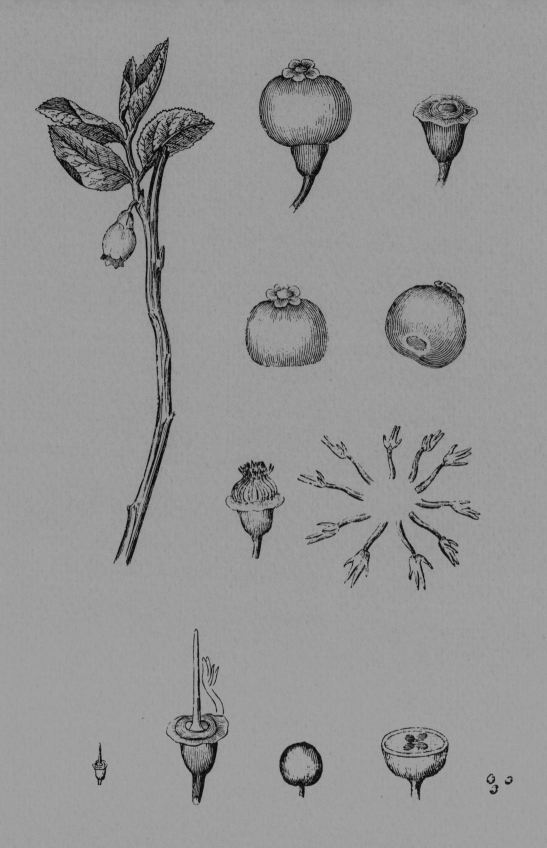

BLUEBERRY

Bertha Nathan, of Maryland, and Blanche Keysner, of West Virginia, gave me almost identical recipes, which they had from their forebears. In Mrs. Nathan's recipe the liquid was strained after twenty-four hours, and in the recipe of Blanche Keysner it was strained after four or five days.

II

Here is a simpler recipe for any amount of berries you can gather.

1. Gather as many blackberries as you can find. Let cold water run over them, if they grew near a road. If you buy them, rinse them a few times and put them into a crock.
2. Add water to cover and mash them; let them stand for twenty-four hours.
3. Strain them, squeezing the berries through a cloth. Return the juice to the crock. Put in 3 pounds sugar to each gallon of liquid and let it stand for a few days in a warm place (65°-70°). It should ferment of its own accord, but if it does not, you can add ½ oz. yeast here, in the usual way. Be sure to warm the must a little before adding the yeast.
4. Stir and skim each day.
5. When the fermentation has ended, put the wine in glass jars or gallons and cork them loosely.
6. When fermentation is completely finished, clear, fine, and bottle.

III

Here is another good working recipe.

You need: *4 to 6 lbs. blackberries*
2 lemons
2 gals. water
8 lbs. sugar
½ oz. yeast (2 packages)

1. Wash the berries and put them into a crock.

2. Wash and pare the lemons very thinly. Put the rinds into the crock.
3. Boil the water for about five minutes in an enamel vessel, then pour it into the crock over the berries and the rind. Mash the berries with a wooden spoon. Let the must stand four days, stirring it every morning and evening.
4. On the fifth day, strain it through a cloth, clean the crock, and put the must back into it.
5. Dissolve the sugar thoroughly in the liquid.
6. Add the juice of the lemons.
7. Dissolve the yeast in ½ cup warm water. Heat the must to lukewarm and pour the yeast into the crock. Cover and stand it in a warm place (65°-70°).
8. When fermentation has ceased, after two or three weeks, strain the wine, clear, and bottle.
9. To add a flavor from the Far East, put in a stick of cinnamon and a dozen kernels of black pepper while the wine is fermenting.

If you want a spicy blackberry wine, put in a spice bag consisting of ½ ounce ginger root, six cloves, half a stick of cinnamon, and twelve peppercorns. Put these in a little bag of clean white material, tie with a long string, and hang the bag in the crock during the fermentation period. At first the wine will seem highly spiced, but after a year the spice flavor will be an undertone, and the wine will have an exotic bouquet and taste.

IV

Nearly every one of my friends in Putnam County gave me recipes for blackberry wine. I will repeat the one told to me by Mr. Harkins, my nearest neighbor and the good guardian of my cattle.

1. Get your berries, plenty of 'em.
2. To every four quarts you got, add a quart o' boilin' water.
3. Leave 'em stay covered up for mebbe two days, just runnin' a clean wooden stick 'round 'em now an' then t'mash 'em up.

4. Then run it through a piece o' cloth into a barril.
5. For every gallon you got, you put in two pounds o' sugar. Close it an' let it stay for six months to a year. Then you'll have a wine that'll make you hear the angels singin'. And I say Amen to that! Oh, yes, make four gallons 'stead o' two, 'cause your friends'll be comin' in beggin' for it!

[*Blueberry Wine*]

WHEN I FIRST BOUGHT MY FARM some twenty-odd years ago, you could not walk through the greening fields, so thick were they with blueberry bushes. There they were, in rich abundance, begging to be used.

The first result was pies made by my fair lady. Of course, Rosita the Indefatigable did the picking. If you have eaten a pie made of freshly picked blueberries with the undercrust flavored with cinnamon and first baked a little, and the upper crust also garnished with brown cinnamon powder, then you know why the angels sing in the paradise of gastronomy.

And later on, an even more celestial touch was added to that master work of culinary art. That was when I met Horace E. Hillery, the Putnam County historian.

"Blueberry pie," quoth he. "Here is the kind my mother made in golden-sunned Alaska, and if you can beat that, I will live on dry stone fences for the rest of my days.

"Put fat blueberries in a good-sized crock, just about two inches deep, and put a layer of sugar on 'em. Lay two inches more of plump berries and cover these with sugar. Keep on doing that until the crock is near full. Let it stand the winter. When early spring calls for a delicious tang, a pie with plenty of these berries answers when nothing else would."

"Amen," say I.

These are holy words, if you believe in the Kingdom on earth. If you will follow the recipe to its conclusion, you will say Amen too. P.S.: You need not add any brandy to this. Nature has done a good job.

Now for the wine. Many recipes came to me from friends. On the whole, they are similar.

This is a composite, and the one I use.

You need: 8 qts. blueberries (you can use 6, but 8 is better)
2 gals. water
5 lbs. sugar
½ oz. yeast (2 packages)

1. Pick or buy the blueberries and wash them well.
2. Put the water into an enamel pot and bring it to a boil. Let it boil for a few minutes and then pour it over the berries. Cover and let stand for twenty-four hours, mashing the berries a few times.
3. Strain the juice into an enamel pot, pressing the berries through a cloth until every drop of juice is out. Clean the crock and pour the juice back into it.
4. Add the sugar and dissolve it thoroughly. Heat the liquid to lukewarm.
5. Dissolve the yeast in ½ cup warm water and pour it into the crock. Cover and stand in a warm place (65°-70°) to ferment. This will take from ten to twenty-one days. When fermentation has ceased, strain, clear, and fine. When it is a shining, bright, reddish color, bottle and cork.

A man in Ohio sent me a recipe in which he did not use any

yeast, contending that the berries will do their own fermenting, like grapes. That is sometimes so, but to be on the safe side, use yeast.

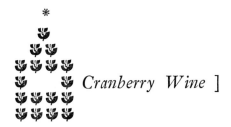

[*Cranberry Wine*]

SINCE CRANBERRIES ARE so intertwined with one of our happiest of festivals, Thanksgiving Day, it is no more than just that we should make a cranberry wine—and a good wine it is.

The North American Indians thought the settlers should have the pleasure of this startlingly red, firm fruit. The Indians had eaten cranberries as far back as they could remember, using them in their pemmican and as a cure for liver troubles.

The Colonists liked the berries so much that they sent ten barrels of them as a gift to Charles II, and I hope he enjoyed them as much as did the Colonists and Indians.

You need: *8 qts. cranberries*
2 gals. water
6 lbs. sugar
3 lbs. raisins
¾ oz. yeast (3 packages)

1. Wash the cranberries and put them into a crock.
2. Boil the water and pour it over the cranberries. Mash the berries thoroughly with a wooden spoon. Then cover them and let them stand for a week, crushing them daily with either your hands or a wooden spoon.
3. On the eighth day, strain the liquid into an enamel pot, squeezing the fruit dry through a thick cloth.

4. Dissolve the sugar in the liquid.
5. Cut up the raisins as well as you can and add them.
6. Heat the liquid to lukewarm, then pour it back into the crock.
7. Dissolve ¾ ounce yeast in ½ cup warm water and add it. Cover, set in a warm place (65°-70°) and let it ferment. Fermentation will take about two or three weeks. Strain into glass jars, let the wine rest for a few days, then decant, fine if necessary, and bottle.

The wine will gain greatly by aging, and so it is wise not to drink it for a year; a longer period is even better.

 Currant Wine]

I HAVE HAD so much difficulty buying and growing currants that I almost decided not to include this recipe. Yet you may be able to get the fruit, or you can buy dried currants in any health store. So here is the recipe for making the wine.

You need: 8 to 10 qts. fresh currants
2 gals. water
6 to 8 lbs. sugar
1 slice toast
½ oz. yeast (2 packages)

1. Wash the currants thoroughly. Put them in a crock and mash them well with a wooden spoon.
2. Boil the water, let it cool, and pour it over the mashed fruit. Let this stand for two days, mashing the berries twice each day.
3. Squeeze through a cloth until the pulp is dry.

CURRANT

4. Return the liquid to an enamel pot and dissolve in it 6 to 8 pounds sugar (depending on how sweet you like the wine). Warm it slightly and return it to the crock.
5. Prepare a slice of toast. Dissolve the yeast in ½ cup warm water. Cover the toast with the yeast and put it, with the excess yeast, into the crock. Set it in a warm place (65°-70°) to ferment. This will take from fourteen to twenty-one days.
6. When the visible fermentation has ended, strain the wine into glass jars or gallons, let it stand a few days to set, then rack and fine, if necessary.

This wine must not be drunk for a year; two or three years is even better.

The recipe I have given can be used half-in-half with raspberries or gooseberries. These are put in with the currants.

Again, honey can be used instead of sugar: a pound of honey for each pound of sugar. This will vary the flavor of the wine pleasantly. Use a honey that is unblended and unboiled.

If you use black instead of red currants, the results will be a rich, heavy wine like a heavy port or a Malaga. But you *must* let this wine age three or four years before drinking it. Then you will enjoy it as much as any fortified after-dinner wine.

Elderberry Wine]

PERHAPS THE MOST VIVID of all homemade wines is the one from this deep purple berry. It makes a strong wine, a wine with pronounced personality, a masculine wine. It was the first wine I made, and it was an experience never to be forgotten. I will tell you of it, for it will be a lesson to you, as it was to me.

Like all who have made a "first" wine, like all who will make their first wine, I was impatient as an ant on the run.

My son, André, Rosita, our faithful houseworker, and I made the wine. Rosita, half South American Indian, short, with long hands and strong body, did all the picking. To make a ten-gallon barrel of wine, you need mighty much picking of berries. My son, a most perfect son, did all the work; and I did all the talking. Friends and neighbors entered into the picture, also talking. There was great excitement everywhere, which I fanned with continuous descriptions of the marvelous wine to come.

For hours and days Rosita brought in half bushels of berries; my most perfect son worked; and I declaimed. The berries were finally pressed, all the steps were followed according to Hoyle, the must was put into the barrel, and the barrel was carried down into the cellar. Many were present to celebrate the occasion.

Then—day after day I went down to the cellar, dozens of times, to see how the wine was getting along.

Now, I am a firm believer in devils and other kinds of evil spirits rampant in the world, and I know one of those devils got into the barrel of wine and was lying in wait for me. Of course, I did not realize this until later.

One day when I went down to look—*bang!* There was a thundering explosion. I had forced the bung down tight, and it shot up to the ceiling and a wild red stream shot out of the opening with a fierce hissing—over me, over the old hand-hewn beams, over the stone floor, over everything. I could swear I saw a red-faced, red-caped, red-tailed devil leaping out of the bunghole hissing fiercely at me, and I believe I smelled sulphur.

The thunderous bang brought the whole family into the cellar —no, only down the steps, for a wide red stream was rushing out of the narrow bunghole. Only my most rare son, André, came all the way down, also to be baptized by the red, fermenting elderberry juice!

We came up and used all kinds of soap and hot water to get the red stains out. It was tough work!

When I told Mr. Harkins, my good neighbor, of my theory of the devil in the barrel, we had a lengthy philosophical discussion about whether it was a true old-fashioned Yorker devil or whether it was a concrete living image of my impatient personality in the form of a devil. More than twenty years have passed, and now and again we still argue that point. (At this writing, Mr. Harkins has just passed his seventy-ninth birthday; he still rises every morning at 4:30 to attend to the milking of the cows. He dislikes new-fangled electric machines: " 'Taint just nat'ral." And I agree with him!)

And now it is the beginning of August, the month when you must begin to pick the elderberries for making wine. You must work fast, otherwise the birds will get them before you do, and I am certain they have just as much pleasure eating them as you or I have making wine of them.

It is also the time for picking and drying slender-leaved, scented tarragon, and the room in my old farmhouse is drenched with the delicate, delicious, herbal fragrance. The scented air pervades the old wooden beams and plastered walls and shelves of books.

There is an additional pleasure in making wine in these hundred-year-old spicy, square rooms where I have spent many years. I am the second owner of this house and farm. The Coles acquired the land in 1730. The original house was a way up the road, and burned down. The old stone, dirt-floored cellar is still there. The present house is a rambling place where square rooms were added when there was need of them. But I had better return to my sheep.

The place was overrun by elderberry bushes when we first came to the farm. They were everywhere, even as they were and are all over the world. They are said to hold one of the three most prominent places in folklore.

There is no end to the customs connected with the elderberry. The "Elder Mother" appears again and again in popular traditions and customs. Every land has green tales in varying lights and shades, divided between chilling, gloomy arts and homey, healing scenes.

In medieval days, a loving mother among the Nordics put a necklace of bits of elder on a thread around the throat of her child to prevent the dreaded, painful "teething fit."

Out in blond Denmark, folks held the lovely idea that if you stood at midnight under the elderberry bush, you would see the King of the Elves in all his glory, with his glittering attendants in procession, including the Elder Mother.

Here, too, the berry was a cure for the dreaded pains of the teeth, but here it was accomplished by placing a twig of the elderbush in the ground, in absolute silence.

The wood of the elder must not be used for a cradle unless you first ask permission of the Elder Mother. If you don't, she will turn and twist the cradle crazily in all directions and even strangle the child! Nor can the bush be cut down for any purpose unless permission is asked first of the Elder Mother, and apologies are made for the cruel act.

That custom of apologizing to trees or animals before destroying them is found the world over. I came across it again and again when collecting tales in Latin America.*

In England, the elderbush is the friend of man and is planted near the house to ward off evil, especially to ward off lightning, for lightning cannot touch it.

The bush has also had an important place in medicine. The Ebers Papyrus of ancient Egypt mentions it, and Hippocrates speaks

---

* See *The King of the Mountains: a Treasury of Latin American Folk Stories*. Vanguard Press, New York.

of it. In the days of Henry VIII, it was used as a remedy against the plague.

In our United States, elder-bark tea was popular as a laxative and a diuretic. The leaves were used as a poultice for inflammations and, steeped in hot water, as a gargle for an inflamed throat.

Children made flutes from the young elder twigs, and men used the hollowed, thicker branches as spouts for catching the flowing maple syrup. In the mountains of Kentucky, there is a poetic belief: If you go on Christmas Eve into the woods where elderberries grow, you will see the lovely, lacy, sweet-scented elder blossoms blooming in the white snow.

To me, such stories woven out of the fabric of mind, air, sun, sky, woods, and fruits are part of the wine.

The poet Leigh Hunt tells of an English club called "The Elders," consisting of men who met for the sole purpose of drinking elderberry wine. They drank the rich wine made by club members and kept barrels for years to improve the taste. I, too, have made it for years; and it is one of my favorite wines, with its opulent Malaga-ish and rich, heavy, port-ish taste, and here is how you can make it and enjoy it too.

The Recipes

In August, the clusters of black-blue elderberries are ready to drop off, or to be picked by you or eaten by the birds. If you don't grow them yourself, you will find plenty of them along country roads, and the owners of the land, I am sure, will give you permission to pick them, even as they have given it to me. Gather them in a plastic pail or paper bag, and you are ready.

Now, there are more than one hundred and one ways of making elderberry wine. I shall give you but a few recipes. You can find others in olden books, or you can make your own variations.

I

I will first set down the simplest way of making it.

You need: 2 gals. water
8 qts. elderberries
6 lbs. sugar
½ oz. yeast (2 packages)

1. Put the water into an enamel pot.
2. Add the washed elderberries.
3. Bring to a boil and boil for twenty-five to thirty minutes.
4. Strain the berries, squeezing every drop of juice from them, and put the liquid into a crock.
5. Dissolve the sugar completely.
6. When the liquid is lukewarm, dissolve the yeast in ½ cup warm water and pour it into the crock. Cover, put it in a warm place (65°–70°), and let it ferment. This will take from ten to twenty-one days.
7. When the wine has quieted down, strain into jars and let it clear. If it does not clear of its own accord, fine it with eggshells or isinglass and then bottle and cork.

II

You need: 8 qts. elderberries
2 gals. water
6 lbs. sugar
1 lb. raisins
1 oz. fresh ginger root
6 cloves (or any spices you prefer)
3 lemons
¾ oz. yeast (3 packages)

1. Wash the berries and put them into a crock.
2. Boil the water in an enamel pot and pour it over the berries.
3. Let them steep for twenty-four hours, and then mash them with

a wooden spoon or with your hands until you have a mushy pulp.
4. Strain well through a cloth into an enamel pot. Get every bit of juice out.
5. Dissolve the sugar thoroughly in the liquid. Add the raisins, cut up, the spices, and the thinly pared lemon rinds. (The spices can be put into a little bag and suspended in the liquid.)
6. Bring the liquid to a boil and let it simmer for at least one hour. Then strain it into a crock.
7. Add the juice of the lemons.
8. Add the yeast, dissolved in ½ cup warm water, when the liquid is lukewarm. Cover and proceed as in the first recipe.
9. Strain the wine into jars. Clear and bottle.

Most important of all, let the wine rest for at least two years before drinking it. The older it is, the better; and it can be as fine as any port or Malaga or Madeira.

This is the recipe I use in my own wine making.

III

A. You can vary the wine by adding about a dozen dried prunes just before you add the yeast. This will give an even richer taste and a brownish flavor.
B. If you want to make a stronger wine—say up to 18 percent alcoholic content—without the addition of brandy, add 1 or 2 pounds barley or wheat at step 5.

ELDERBERRY

 Gooseberry Wine]

FOR THOSE WHO GROW gooseberries or wish to buy them, here is a way of making them into a dry, light-colored, tauntingly flavored wine, excellent with plainly cooked fish or chicken or with fish soufflés.

The name "gooseberry" has always intrigued me. A goose is a lovely creature to look at, and it makes a most excellent feastly dish. Why this berry is allied in nomenclature to the goose, I discovered through my botanical angel, Elizabeth C. Hall, the excellent librarian of the New York City Botanical Garden. She informed me that she found two sources for the name: one, a fine folkloristic theory, and the other more scholarly.

Folkloristically, the berries were so named because the good folks of England found that they made an excellent sauce for "green geese"—goslings. The other explanation is that the word is a corruption of the Dutch word *kruisbes* or *gruisbes*. *Kruis* means "cross," and *bes* means "berry"—the berry that is ready to eat just after the festival of the Holy Cross (September 14). There is a third explanation that *kruis* (in the sense of "curling" or "crisp") is a term applied to the tiny hairs of the fruit.

Personally, I shall accept the English explanation, and I hope you will too. It is homey and friendly.

In Sussex, England, the berries are called "goosegogs," as an Irish lady in London informed me. They are much more popular in England than here in North America. There I have seen them in varying shades—greenish, purplish, yellowish, reds—sometimes as large as plums. They are also very popular in England for wine—in fact, the gooseberry is one of the most popular berries for wine making.

The Recipes

I

You can use gooseberries when they are green and underripe or when they are yellowish and soft. I have grown gooseberries, and if you do, I advise you to pick them green. Otherwise the birds will come before you, and all you will find will be green leaves and no berries.

You need: *6 to 8 qts. gooseberries*
4 to 6 lbs. sugar
2 gals. water
¾ oz. yeast (3 packages)

1. Put the gooseberries into a crock and mash them well with a wooden spoon or a wooden mallet.
2. Pour the sugar over them.
3. Boil the water in an enamel pot, and pour it over the fruit. Stir and mash with a circular motion until all the sugar is dissolved. Then let the liquid cool to lukewarm.
4. While it is cooling, dissolve the yeast in ½ cup warm water. Then add it to the crock.
5. Cover; let it ferment, which will take from two to three weeks. Stir once every day with a wooden spoon.
6. When the fermentation stops, strain the wine into glass jars or gallons and let stand from four to fourteen days. If it is not clear by then, clarify with isinglass or eggshells, and then bottle.

Gooseberries are often combined with other berries for wine making—with currants, elderberries, etc. This will give the wine a different color and vary the flavor. If you are adventurous, you can try such combinations.

II

This recipe was given to me by Bertha Nathan. It has been used in her family, down in Maryland, for many years.

You need: *1 peck gooseberries*
1 gal. water
3½ lbs. sugar to each gallon of juice

Let the berries stand in the water overnight. To get the juice, the berries must be mashed thoroughly. Strain the next day. To every gallon of juice add 3½ pounds sugar.

I believe the must would ferment without yeast. But perhaps the good housewives who made the wine believed that the use of yeast in wine making was such common knowledge that it did not have to be set down. Anyway, you can try it, and if in three or four days there is no sign of fermentation, you can dissolve ½ ounce yeast in warm water or must and add it. Thereafter you proceed as usual.

Whichsoever recipe is used, the result will be a very nice white wine, ready to drink after a year or more.

 *Huckleberry Wine*]

YOU MAY PICK HUCKLEBERRIES in the country or buy them in a fruit store. Whichever you use will give you a good red wine, sweet or dry, according to your taste. If dry, it is somewhat in the class of a light Bordeaux, or if sweet, a light port.

Huckleberries were a popular fruit among the North American Indians. Large groups went out into the swamps and along the streams to pick the berries when they were in season. They cooked them and brewed a medicine to cleanse the blood.

The wine is made in the same way as other small-berry wines.

You need: *8 qts. huckleberries*
2 gals. water
4 to 8 lbs. sugar

1. Get huckleberries, "boughten" (if so, wash them and let them dry) or picked; put them into a crock and crush them well with a wooden spoon.
2. Pour the water over the fruit and in it dissolve the sugar—use 6 or even 8 pounds if you want a sweet wine. The must should start to ferment quickly. Stir it a few times each day.
3. On the third day, strain the juice and put it back into the crock. (If fermentation has not yet started, you can help it with ½ ounce yeast dissolved in ½ cup warm water. If you do, it would be well for you to heat the must a little, since yeast starts much better in warm liquid.)
4. When the visible active fermentation has ended, strain into glass jars and let the wine rest for two weeks or more. It should be clear by then. If it is not, fine it. When dark ruby clear, bottle it. This wine, like most, should stand at least a year before you drink it.

[*Mulberry Wine*]

I WILL INCLUDE a mulberry wine in the book, though I know it is not easy to get these berries unless you grow them in your own place in the country or find them on some old deserted farm. In back of my rose garden I planted two weeping mulberry trees, which I bought from a nursery in Maryland after much searching; and for many years I, and the birds, have found them a great delicacy.

One summer I put green, vapid-looking cloth snakes over the hanging branches to keep the birds away. They stayed away, and I stayed lonely, which gave me time to realize what a clotted, cowardly deed I had done. I apologized to the birds and promised never to do this again. And I haven't! Since then we have both shared the tender, delicious, deep-red-black berries; and there is, oft as not, enough left to make into wine.

Besides finding pleasure in the mouth-watering fresh flavor of the berries, I have a sentimental and nostalgic love for them. I ate them in abundance in my childhood, and always associate them with the second Shakespeare play I saw, the Ovid tale of *A Midsummer Night's Dream*, one of my three favorites of the poet's works. I saw it at the age of eight and wept over the sad death of Pyramus and Thisbe under the mulberry tree, for the tragedy was real to me then. Did you know that until that tragic incident all mulberries were white, but after sweet Thisbe and valiant Pyramus soaked the ground under that mulberry tree with their innocent blood, the berries turned purplish-black in color?

If you like Chinese thoughts and literature as I do, you will find the mulberry even nearer to your heart for the praises sung of it in the torn land of graceful jade. Read the lines of Pao-P'u-Tsu, which can be found in the beautifully printed book, *Fragrance from a Chinese Garden,* by Alfred Koehn, and judge for yourself.

". . . Take the ashes of the Mulberry tree and the gum from the trunk of the Peach, mix them together, and by drinking this you will be cured of one hundred different illnesses. When you have taken this for a long time, you will never require food and your body will emanate light."

If I did not love food as much as I do and were not exposed to the most delicious food possible, I should mix those two substances and drink them. But I maintain that "requiring" food is one of the most laudable of virtues and one of the great joys of life, so I will not burn down either of my two weeping mulberry trees.

Of course, in China, for centuries, mulberries have been of utmost importance as an ideal food for silkworms and as an ideal inspiration for poets and scholars.

There is a well-wrought folktale told in China about the fruit. In the Tse dynasty there lived a man named Chang Ching. On a moonlit night he walked into his garden, and there in a corner near his house stood a tall, slender lady. She motioned him to approach, and when he was near, she said:

"This place where I stand is your honorable person's mulberry ground, and I am a shen [a fairy]. If you will make, next year, in the middle of the first moon, some thick congee [mush] and give it to me, I will undertake to make your mulberry trees a hundred times more fruitful."

The good man made the thickest congee of rice that could be made and gave it to the fairy, and for it the silkworms in his mulberry grounds increased a hundredfold. Since then, Chinese folks eat a very thick rice congee on the fifteenth day of the first month of the year.

Pliny called the mulberry the "wise tree," for it has the wisdom to blossom late in the spring when no cold can hurt it.

How contrary people are! In Burma the tree is worshiped, whereas in Europe the tale runs that the Devil blackens his boots with the juice of the berry. And what is wrong with that! What is wrong with the Devil?—but that is for another book. Instead, I would like to hear of anyone who has not danced to the words and tune of the old English folksong: "Here We Go Round the Mulberry Bush"! I am sure there are not many.

Since mulberries are flavorsome not only to silkworms but to humans, it is just to make a good wine of them, and here is how it is done.

The Recipes

I will set down two recipes for making mulberry wine; both are very simple. Let me say here that the more berries you use, the richer the wine will be in color. It will range from a rose to a deep red, and in flavor from a dry to a heavy port wine type.

I

You need: *2 gals. water*
2 to 4 lbs. sugar
4 to 8 lbs. mulberries
4 lemons
1 slice toast
½ oz. yeast (2 packages)

1. Put the water in an enamel vessel with 2 to 4 pounds sugar (or even a larger amount if you want a very sweet wine) and dissolve it thoroughly.
2. Now add the ripe mulberries.
3. Wash and pare the lemons very thinly and put the rinds into the pot. Then add the juice of the fruit.

4. Boil this for thirty minutes, then pour into a crock.
5. Let it cool off while you prepare a slice of toast and the yeast, dissolved in ½ cup warm water. When the liquid is lukewarm, pour the yeast on both sides of the toast and put both into the crock.
6. Cover, let stand in a warm place (65°-70°), and fermentation will begin quickly. This will take from seven to fourteen days. Strain into an enamel pot, clean the crock, and pour the must back into it.
7. Cover again and let stand for another week or more until all fermentation has stopped. Then strain the wine into glass jars and let it rest for two weeks or more. If it is not clear by then, rack or fine it until perfectly clear, then bottle and cork.

Let the wine stand at least two years. The older the wine, the better it will be.

II

You need: *2 gals. water*
4 lbs. mulberries (or more)
3 to 6 lbs. sugar
1 oz. fresh ginger root
2 to 3 cloves
1 slice toast
½ oz. yeast (2 packages)

1. Put the mulberries into the water. Mash the berries well with a wooden spoon. Bring to a boil and simmer for thirty minutes. Then let stand for twenty-four hours, well covered.
2. Strain through a thick cloth into an enamel vessel. Put in the sugar (the amount depending on the degree of sweetness you want) and dissolve it thoroughly.
3. Heat again, adding the spices. Simmer for another twenty to thirty minutes, then strain and pour into a crock.

4. Proceed as in the first recipe.

This wine should not be drunk for at least a year.

[Raspberry Wine]

ONE OF THE MOST FRAGRANT and popular wines with me and my friends is raspberry wine. It is pleasantly and delicately dry, with the sweetness and perfume and flavor of the berry.

The North American Indians were very fond of these berries and had a gracious and grateful ceremony connected with the first eating of them each year. They prayed to the "spirit" of the berry to help them in their undertakings in peace and war. Then they prayed for forgiveness for eating it. They also used the leaves for tea, even as we do to this very day.

Like all fruit, the raspberry has folk meanings and folk uses in different parts of the world. Filipinos hang raspberry vines on the doors of huts where there is a dead person, so that when the spirit tries to return to the house, it will become entangled in the vines.

British and American folks made a raspberry vinegar to cure sore throats and colds. The "raspberry" or "Bronx cheer" in the U.S.A. is a slang expression of defiance and contempt. To express derision, "Give him the raspberry" is often heard at wrestling matches and prize fights.

But to dream of raspberries is a happy omen indeed. It means all kinds of good tidings: fidelity on the part of one who loves you;

good news from far lands; happiness in marriage; success in undertakings; and everything else that brings good cheer.

As a drink (and as a food), the berry has been in favor for many years. Raspberry gin was much in demand in England in the eighteenth century.

The Recipes

For the wine, it is best to use homegrown berries. If you buy them, be sure to let cold water run gently over them for a long time, without bruising them. Here are two recipes; either of them will give good results. I have found it best to use a quart of berries to a quart of water, but this need not be a hard-and-fast rule. If you use less, you will get a good wine, but it will just lack the richness the additional berries give.

I

You need: 4 to 8 qts. raspberries
2 gals. water
6 to 8 lbs. sugar
½ oz. yeast (2 packages)

1. Put the berries into a crock.
2. Bring the water to a boil and pour it over the berries.
3. Cover and let stand from two to three days. Stir and mash once or twice each day.
4. Strain the liquid through a cloth into an enamel vessel, making sure that every drop of juice is squeezed out of the berries.
5. Add the sugar, the amount depending on how sweet you want the wine to be. Be sure to dissolve it well.
6. Heat the liquid and pour it back into the crock.
7. Dissolve the yeast in ½ cup lukewarm water and pour it on the liquid. Cover and set in a warm place (65°-70°) to ferment. This will take from ten to twenty-one days.

RASPBERRY

8. When fermentation has ceased, strain the wine into glass jars and let it rest for two or more weeks. Then siphon off into bottles any of the wine that is completely clear. The rest—or all of it if it is cloudy—fine with eggshells or isinglass, and bottle.

The wine can be drunk one year later.

II

Raspberry wine can be made without yeast. Just keep in mind that if the recipe does not work out satisfactorily, you can always add the yeast.

This method has been used since 1820 and probably much before; many fine wines have been made in this way. It is almost the same as the first recipe, with the following slight differences. It is suggested that either cold or lukewarm water—not boiling—be poured over the berries. Then allow the berries and water to stand for forty-eight hours, mashing them a few times a day. After the mash is strained, set it in a warm place to ferment. (If fermentation does not start, add ½ ounce yeast.) When the fermentation has ended, strain the wine into glass jars. Rest it for two or more weeks, and if it is not perfectly clear, fine it.

Whichever way the wine is made, it should not be touched for one year. Two is better—the older the wine, the better it is. I have a raspberry wine seventeen years old that was given to me by Bertha Nathan, of Maryland, of whom I have spoken in other parts of this book. It is perfect in color and ambrosial in taste and bouquet, and not only warms the cockles of the heart but the bones and flesh and veins and arteries of the body as well.

Raspberries have been used in flavoring brandies, vodka, and gin, but of this I will speak in the chapters on Unusual Drinks and Flowered, Fruited, and Flavored Brandies.

 Strawberry Wine]

STRAWBERRIES YIELD A DELICATE, fruity, and refreshing wine, well worth the making. If you don't have a strawberry patch in your garden, or they are out of season, you might try frozen berries or strawberry jam. I have been told that a good wine can be made from the latter, but I have never tried it. I always manage to get the fresh fruit.

Should any strawberries drop when you pick them in your patch, let them lie on the ground. They must be left there for the poor, who have no berry patches. Such was the custom in the olden days, and a gracious custom it was. And when old folks carrying the fruit passed a church, three berries had to be offered to the church.

This was the fruit sacred to Freya, the Norse goddess. There are a large number of legendary tales and customs centered around it in the Northlands. When these people were converted to Christianity, the berry was dedicated to the Virgin Mary, and it was a sin to eat the fruit before Saint John's Day. Particularly so, since it was a symbol of the children who had died young and ascended to Heaven hidden in the berries. Saint John the Baptist is said to have lived on the fruit, and for that reason, Don John, the son of John I of Portugal, who was devoted to the Saint, took the strawberry as his heraldic device.

In our own country, some folks still use the leaves in teas for sore throats and for jaundice. It was also a cure for liver trouble. How much less expensive than our modern drugs and in many instances quite as efficient, provided time was allowed as part of the medicine—a perfect cure for so many ailments.

For the North American Indians, strawberries were an important fruit. The women and young people gathered them with song, and the picking was a pleasurable social function. The Iroquois had a strawberry thanksgiving festival because it was the first fruit of the season to be gathered.

The Chippewa Indians called it "odamin" (heart berry), and they have an amusing tale of the fruit.

Once there was a husband who had many quarrels with his wife. One day she went away. But the man really loved his wife and wanted her in his wigwam. He became very sad. The Great Spirit saw this and felt sorry for the man and wanted to help him.

"Do you really want your wife back?" he asked.

"I do."

"I will bring her back to you."

The Great Spirit went away seeking her and soon saw her running through the woods. He ran after her. She saw him and knew why he pursued her, so she ran faster.

The Great Spirit tried to stop her. First he threw before her a basket of blueberries. She did not look at them; she just kept on running. He threw cherries in her path, and he threw every kind of fruit in her path, but she would not stop.

"I will try one more fruit," the Great Spirit said, "a fruit I have just created."

So he threw in her path the new berry—it was the strawberry. She stopped to look at the rose-bright, glistening berry. She had never seen such a berry before and, like all women, she was curious.

She bent down to pick one up and then put it in her mouth. It was sweet. She had never tasted a berry like it, and it was very good. She sat down to eat more berries, and so the husband, who had been looking for her, found her.

"Come and eat this fine new fruit," she said to him.

He sat by her side and they ate the fruit together in peace. The woman did not run away again, and the man did not quarrel, so the two lived together happily ever after.

With this happy note, the wine is in order.

The Recipes

There are different ways of making this wine, and I will set down two for you.

I

You need: *strawberries—any amount, depending upon how much wine you want*
sugar—about ½ pound for each quart of berries
a pint of strong liquor, sweetened

1. Get as many quarts of strawberries as you wish. Put them into a crock. Add the sugar. Cover and let stand for a full day, macerating and squeezing the berries a few times as hard as you can.
2. Strain, squeezing every drop of juice out of the berries.
3. Cover well and stand in a warm place (65°-70°). The must should ferment without yeast. If it does not, put in ½ ounce yeast (2 packages). But I am quite certain this will not be necessary.
4. Let ferment for about four days, then add any strong, sweetened liquor (the least expensive)—gin, vodka, or alcohol. This will set the color and the flavor of the berries, both of which disappear very easily.
5. Now let the must run its full course of fermentation. When the wine is completely quiet, strain it. Let it stand for two more weeks, and if it is not absolutely clear by then, fine.
6. It is advisable to let the wine rest, say in gallon bottles, for another month before the final bottling and corking.

II

You need: 8 qts. strawberries
2 gals. water
6 to 8 lbs. sugar
½ lb. raisins
2 lemons
¾ oz. yeast (3 packages)

1. Wash the strawberries under cold water if you bought them, and put them into a crock. Mash them thoroughly with a wooden spoon.
2. Add the water and let stand for three days, mashing and stirring the berries at least three times a day.
3. Strain the liquid through a kitchen towel, squeezing the juice out of the berries.
4. Add 6 to 8 pounds sugar, depending on how sweet you want the wine to be, and dissolve it thoroughly.
5. Add the cut-up raisins and the thin rinds and juice of the lemons.
6. Heat half the liquid and return it to the cold, so that the whole becomes lukewarm.
7. Dissolve the yeast in ½ cup warm water and put that in. Cover and let it ferment.
8. When the fermentation is ended, strain the wine into gallon jars and let it rest for two weeks.
9. Then siphon into bottles if clear; if not, fine it.

In both recipes, the wine should mature for at least one year.

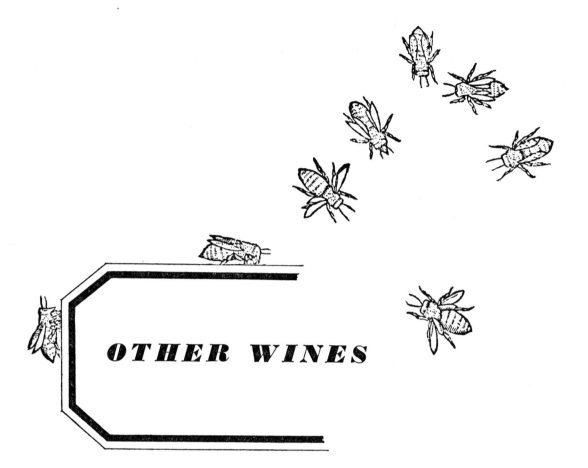

OTHER WINES

ALMOND

 *Almond Wine*]

"... AND BEHOLD, the rod of Aaron for the house of Levi was budded, and brought forth buds, and bloomed blossoms, and yielded almonds."

An almond wine! What a rich, Biblical-bannered wine! There could not be a more lovely, heavenly herald of spring than the fragrant blossoms on the bare brown branches of the almond tree. The annual reincarnation of life!

Sweet gifts to man are "a little honey, spices, and myrrh, nuts, and almonds." These are the chanting words in Genesis when Israel sent a gift of almonds as a sign of love and friendship. The wood of the almond tree, a blossom of the branches, the fruit to be gathered, all were part of the Hebraic ritual in the tabernacle and in daily life.

The ancient Phrygians, seeing those blossoms on the bare

branches while all green life was still cold and barren, believed the tree was the father of all life. For it was the beginning of all green growth.

Their god, Attis, who was loved by Cybele, the mother of the gods, was born of Narra. She conceived the fair young shepherd by putting a ripe almond in her bosom.

This scented symbol of new hope is also in the lore of other lands. In the Mohammedan world the almond is the mark of heavenly hope. It brings assurance, warmth, and food when other trees and fruits are still nascent.

The great genius of the Greeks built a comely tale around the almond tree, delicate as the spring blossoms.

Demophon was on his way home from the long siege of Troy. His oared ship, battered and beaten, was wrecked upon the shores of Thrace. There the handsome Greek (they were all strong and handsome in that war) met the king's daughter, And, as was the way of those Greeks, he fell in love with her and she with him, and he promised to marry her.

But first he must go to his own land of friends and kin to bid them farewell and to tend the affairs of home. He went and, as often happens, forgot the lovely waiting bride.

Faithfully and patiently she waited and waited for the crowning return of her lover. It was a long wait. Too long for love and desire, and with heartbreaking sorrow she pined away and died of the starvation of longing.

Such young-eyed death was too sad even for the gods, and so, for her great, loving, hopeful constancy, they turned her into the universal tree of hope—the almond tree.

But hard-headed Pliny had a different idea about the almond tree. He declaimed that chewing almonds was a peerless remedy for drunkenness.

In Tuscany, the practical Romans used an almond branch as an infallible divining rod for seeking out buried treasures. A good

Etruscan would walk along the field where his dream had told him there lay a buried treasure, an almond branch in his hand, stopping now and then to see which way the twig would turn. The direction in which it turned was the spot of rich fortune for the seeker. There he would dig—with the same result that man has always had when digging for treasures.

Popes used almond wood for their staffs. And I would have you know that almonds have been used for thousands of years, and still are used, in powders, creams, and lotions to enhance the beauty of fair ladies.

Before ending my words of praise for this gracile tree and its blossoms and fruit, I must add one more of its pleasures that is of interest to me and that I am sure will interest you as well, for all the world loves fine food.

Almonds have played an important part in enhancing the pleasure of eating. The Naxian almonds were famous in Greece and were used for dessert. Wheresoever the almond grows, it has been used in many ways. Small wonder, for almonds add a lingering flavor and perfume to food unlike that of any other nut. The Greeks, the East, all the world had, and still have, endless recipes in which almonds play an important part. An almond and lettuce soup, and an almond and cucumber soup—fit for Olympus—are made in mine own home.

With this multitude of recommendations, it should be no surprise to tell you that you can make a delicious wine from almonds. Here is how it is done.

You need: 2¼ gals. *water*
 2 lbs. *raisins*
 3 to 4 oz. *almonds*
 6 lbs. *sugar* (7 lbs. *if you want a very sweet wine*)
 6 *lemons*
 1 oz. *almonds* (*additional*)
 1 *slice toast*
 ½ oz. *yeast* (2 *packages*)

(In this, as well as in practically all the recipes, you can vary the amounts of the ingredients slightly to suit your taste.)

1. Put the water into an unchipped enamel vessel.
2. Chop up the raisins and almonds (but not the extra ounce) as small as you can and put them into the water. Set on the fire, bring it to a boil, and let it simmer for a full hour.
3. As it heats up, put in the sugar and dissolve it thoroughly by stirring with a wooden spoon.
4. Wash the lemons well. Peel the rind off very thinly and put that into the boiling water. Do not put in the lemon juice.
5. When the boiling is finished, strain off the juice into your crock and put into it the juice of the lemons and the extra ounce of almonds, cut in halves. Let the liquid stand until it is lukewarm, when it is ready for the yeast.
6. Prepare a slice of toast and dissolve the yeast in ½ cup warm water. Pour the liquid yeast on both sides of the toast and put it into the crock. Then pour the remaining yeast over all.
7. Cover the crock well and put it in a warmish place. The fermentation will take ten to fifteen days or even longer.
8. When the strong fermentation has ended, strain the wine off into gallon jars (with wide mouths). Let it stand for a week or two. It may be perfectly clear then. If it is not, fine it with eggshells or isinglass.
9. When perfectly clear, siphon off the wine into bottles, cork it lightly, and let it set, watching to see if there is any more visible fermentation. When it is perfectly clear, cork the wine tightly and put it away. Do not drink it for at least one year. The longer you wait, the better it will be.

 Bees Wine]

I ASKED AND WROTE far and wide—I could say to almost every country of the world—for folk wine recipes, and the responses were generous and fascinating. Thus I corresponded with Mr. W. Springle, of England. He is an Army man, deeply interested in the British Army and in English folklore, and he came up with several recipes, handed down from his own parents. One in particular I had never heard about: bees wine.

In his home, when he was a youngster, wine making was as popular as jam making. Says Mr. Springle:

"And during the Christmas holiday, the postman and the milkman and the barber and other friends, all were treated generously by all, and they often ended up a little tipsy.

"There was one kind of wine, bees wine. This was some type of animal life that was kept in liquid in glass jars and fed on sugar. The animals worked all the time, going up and down. The liquid was drained off periodically, strained, preserved and bottled. A rather pungent-tasting liquid which I never liked."

There was a mystery! A wine full of living animals! I was determined to learn something about this wine with "animal life."

When I was in England, I asked high and low, but no one knew anything about it. One elderly couple in Scotland remembered it, but did not know how it was made.

Finally, I went to the National Federation of Women's Institutes, and there Mrs. I. L. Currie, Assistant Secretary, came to my rescue. She contacted the Department of Agriculture and Horticulture of the University of Bristol, and Dr. P. W. Beech of the Research Station gave me the tale of bees wine. Subsequently, I met some folks who knew how it was made. Here is Dr. Beech's scientific description. It is interesting.

" 'Bees wine' . . . is produced from a mixture of yeasts and lactic acid bacteria. The bacteria produce a small amount of mucous material which traps yeast cells—hence the tendency to form masses of organisms that look like rice grains. In the presence of sugar and nutrients (supplied by raisins, juice, etc.) the yeast forms alcohol and carbon dioxide. Some of the gas is trapped inside each grain and causes it to rise to the surface, where, having no hydrostatic pressure, the gas can escape and the grain falls to the bottom once more. The cycle is repeated until most of the sugar is exhausted."

Mr. Springle's living, moving "animals" were the grains going up and then falling down.

Here is the actual process by which bees wine can be made.

You need:
1 gal. water
2 oranges, sliced
½ lb. raisins, cut up
2 tbsps. yeast culture (½ oz. yeast can be substituted)
1½ cups sugar (approximately)

1. Put a gallon of boiled water (spring or deep well water need not be boiled) into a crock.
2. Add the oranges.
3. Add the raisins.
4. Add the yeast.
5. Every day add 2 tablespoons sugar. Do this for at least ten days. When the sugar is all absorbed and the fermentation (the animal action) is ended, filter, fine, and bottle. This wine can be drunk within six months.

I have made it and found it pleasant.

 Brandy Wine]

I FIRST HEARD of this kind of wine in the loveliest village in all Europe, with the loveliest flower-bedecked thatched-roof cottages, and the most beguiling people I have ever visited—Irish, of course—in Adair, Ireland. At the hotel I met an English gentleman. We talked of food and wines, and he told me of a brandy wine he had made.

You need: *4 good-sized potatoes*
 2 lbs. raisins
 2 lbs. coarse barley
 2 gals. water
 8 lbs. sugar
 ½ oz. yeast (2 packages)
 1 qt. brandy

1. Put into a crock the potatoes, well washed but not peeled; the raisins, cut up as finely as you can manage; and the coarsest barley you can buy.
2. Bring the water to a boil in an enamel pot. When it begins to boil, add the sugar and let it dissolve, then pour the boiling sugar-water into the crock.
3. When it is lukewarm, dissolve the yeast in ½ cup warm water and add it. Cover and stand in a warm place (65°-70°). Let the must ferment for three weeks.
4. At the end of that time strain the wine into glass jars. Let it stand another three weeks. Siphon off the clear wine. Do this two or

three times at two- or three-week intervals. It should be perfectly clear. If not, fine and rack it.

5. To the clear wine add a fair amount (1 quart) of brandy, and bottle. Do not drink it for at least one year. The longer it matures, the better.

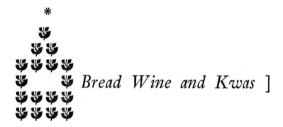 *Bread Wine and Kwas*]

IF YOU ARE ADVENTUROUS in your viniculture, and if you have made it a truly exciting avocation or hobby, make a wine of the Staff of Life.

A wine of bread, a semi-wine I would call it, is very common among Slavic people. I have spoken earlier of kwas. It is a wine, a very mild one, containing about 6 percent alcohol. It is very common in Russia, parts of Austria, and Rumania—in fact, in all countries that have a Slavic population.

To make it pleasing for your use, you will have to add spices or honey—unless you develop a taste for its particular sour flavor.

Of course, the lore of bread runs through the world without end. It was a food from the earliest days of time. There are innumerable miracles connected with it. There are innumerable references to it in all the lore and history of the world. But it is far too great a lore to set down, so I will turn to the wine.

You need: *1 to 2 gals. water*
            2 lemons
            raisins

spices (different kinds, but not too much)
1 to 2 lbs. sugar for each gallon of water
2 to 3 lbs. black bread for each gallon of water

1. Put the water into an enamel pot.
2. Wash and peel the lemons thinly, and put the rinds into the water. Squeeze the juice of the lemons and put that in. Add the raisins, cut up as well as possible. Put the spices (cloves, caraway, coriander, a few peppercorns, and any others you favor) into a little cloth bag and put it in the pot. Bring the water to the boiling point.
3. Dissolve the sugar in the boiling water.
4. Slice the bread and toast it, taking care not to burn it. Put it into the crock. (If you have stale, hard, whole-wheat bread, it can be used without toasting.)
5. When the water is still quite hot, pour it over the bread. Cover the crock and set it in a warm place. If you have used good wheat bread, it will start fermenting very quickly without any yeast. Stir it with a wooden spoon every day.
6. When it has ceased fermenting, let it rest for a few days so that the bread settles. Then siphon off the wine and clear it.

It is best to put the wine into strong bottles—champagne or heavy Burgundy bottles—then cork it, wiring the corks. You may get a powerful, bubbling champagne, which will force out any ordinary cork or even explode an ordinary bottle.

Bread wine matures quickly, but it is best to let it rest for a year.

Kwas

There are many ways of making kwas. The method varies with the locality. In Bukowina, a province of Austria where there are many Slavic folks, kwas was made with apples and had a pleasant cidery, slightly sourish taste.

I have chosen the simplest of the recipes, and you can try it, making it once for the sheer novelty of it. It is modified from a recipe of Harry Rubin and Vasily Le Gros, of the Monastery of Our Lady of Kursk, about a mile from my farm. The kwas is made at the monastery by one of the monks.

You need: *3 lbs. stale, well-baked rye bread*
5 gals. water
3 lbs. raisins
2 lbs. dark molasses (or honey)
½ oz. yeast (2 packages)
1 tbsp. whole-wheat flour

1. Cut the bread into small pieces and put them into a crock or barrel.
2. Boil the water and pour it over the bread. Add the cut-up raisins. Cover the crock well with a tablecloth and let the liquid stand until it cools.
3. Filter it through a napkin or towel, but do not squeeze it.
4. Pour into the liquid the molasses (or honey); use a greater amount if you want a sweet wine. Mix thoroughly.
5. Dissolve the yeast in ½ cup warm water and pour it in, and also add the flour.
6. Cover and place in a warm room (65°-70°). Let the must stand until it starts fermenting, then filter it. Pour it into bottles, putting two raisins into each bottle. After a few days, it should be good to drink.

At the monastery, the priest makes it somewhat differently, using little syrup and no raisins. The result is a very sour drink.

In Bukowina, small whole apples were put in the water before boiling it, and one was put into each glass of kwas when you bought it.

RYE

Cider Wine and Cider Champagne]

YOU CAN MAKE a very pleasant cider wine in a very simple way. When served cold, it has a lovely autumn tang.

You need: 3 gals. apple juice, unpasteurized
3 lbs. honey (or 3 lbs. sugar)
1 qt. brandy

1. Get cider to which NO preservatives have been added, and put it into a crock.
2. For every gallon of cider, add 1 pound honey or 1 pound sugar.
3. Cover and stand in a warm place (65°-70°). It will soon begin to ferment. When the fermentation has ended, decant and clear. This wine will be quite clear and golden yellow. At this point you can add a pint or a quart of brandy.
4. Now bottle, but don't close the corks tightly. Six months later you can seal the bottles, and soon after you will have a fine wine with a cidery tang.

If you like the cider wine spiced, make a spice bag of cloves, cinnamon, and ginger, and suspend it in the wine while it is fermenting. When the fermentation has stopped, the bag should be taken out.

The wine will take on a pink color if you add a few beets during the fermentation period.

It is a good idea to bottle this wine in heavy champagne or Burgundy bottles and to cork it strongly, using a wire basket over the cork. Cider wine has a tricky way of becoming bubbling champagne, and ordinary corks may pop or the bottle may explode.

If you wish it to turn into cider champagne, put a lump of sugar into each bottle. Keep the bottles in a cool place. This is a very pleasant drink and quite popular in England. Of course, it is not a true champagne, but the cork shoots out with a bang, and the wine bubbles, and it has a taste reminiscent of champagne.

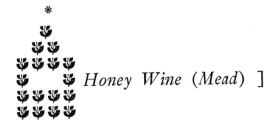 Honey Wine (Mead)]

"CARRY DOWN THE MAN A PRESENT, a little balm, and a little honey, spices, and myrrh, nuts, and almonds."

Thus it says in Genesis, and so will I carry down to you a little help in making good honey wine, or mead.

Like most of us, I associate this wine with the Nordic peoples and with "Old England"—the Anglo-Saxon. In bygone days mead was drunk by the wedding guests for a full month after the ceremony, and thus we have the word "honeymoon." But of course honey wine was known in even older times among all the races in the North and in other lands and climes. The Norsemen drank it during their ritual feasts.

Said Dr. Howells, of Jesus College, Oxford, in the year 1640: "The Druids and Old British bards were wont to carouse thereof before they entered into speculations."

In the early Welsh saga, the Mabinogion, the only drink mentioned is mead.

In the *Book of Taliesin* we read:

"From the mead-horn—the pure and shining liquor,
   Which the bees provide but do not enjoy,
Mead distilled I praise . . .
   God made it for man, for his happiness;
The fierce and the humble both enjoy it."

The basic ingredient of mead is honey, and it has a lore and history that would take volumes to tell. I will give you just samplings in the luminous life of the sweetness of nature.

As early as the first century A.D., Dioscorides speaks of special aromatic oils containing honey and used for anointing. Also, of its use in religious ceremonies and for embalming bodies, in which "fragrant" honey was one of the important ingredients.

Here is a sample of the fantastic flights of imagination in the medieval years. This from *The Boke of Secretes* of the pseudo-magician Albertus Magnus: "The great toe of the right foot of the man anointed with honey or oil or ashes of a weasel will provoke vinery in one thought to be dead in these matters."

To preserve beauty or sagging skin in former years, a lady did not go to a beauty parlor, but she bought or made a mixture of "new laid eggs, rose, honey, and seeds of quince." At least it did not cost as much as the frantic modern striving for the same purpose.

In East Africa, the Masai permit honey wine to be brewed only by a man and a woman set apart for the particular task, and they cannot leave the work until it is finished. For two days before, they must be chaste and must remain so during the work and until the wine is ready for drinking. If they break this custom—no, this law —the wine will be spoiled and the bees will fly away.

Volumes can be filled with the uses of honey as food and as sweetening. In South America, honey was one of the important substances of daily food. People worshiped and prayed to the "Honey Mother"—the deity of the honey—and there are many folktales centered around her.* For centuries honey, not cane sugar, was used to sweeten food and drink.

The Recipes

Before giving the recipes, I want to say a few words about the nomenclature of honey wine, for it has many names; but the fundamental ingredient is the same. It is the additional substances that give the taste and the name. There is mead, or just honey wine, also called hippocras, a word that always gives wings to the imagination! There is metheglin, another musical word; there is sack mead or sack metheglin; there is piment, and there are melomel and hydromel and many more. Each one of these is made somewhat differently, and you will have to decide which kind you like best. Mead can be light or heavy, sweet or dry.

Many of the meads are spiced, and for this you can use ginger, cloves, cinnamon, nutmeg, orange peel, etc. You also can add various flavoring juices of fruits or berries. In all, you can create any kind of mead you like. When spices are added to honey wine, it is called sack metheglin.

You can also flavor mead with elder-flowers, which should be put inside a bag suspended in the must. As for having a greater or smaller alcoholic content, just remember that the more honey you use for a given amount of water, the stronger the mead will be.

I

Here is a simple recipe for sweet mead—honey wine—that I made.

* See *The King of the Mountains: a Treasury of Latin American Folk Stories*. Vanguard Press, New York.

You need: *10 to 12 lbs. honey; any inexpensive honey will do, but try to get one that is unblended and uncooked*
2¼ gals. water
3 lemons
2 oranges
spice bag (containing 12 cloves, 2 sticks cinnamon, ½ oz. bruised ginger)
1 slice toast
½ oz. yeast (2 packages)

1. Put 10 pounds honey (or 12 pounds, if you want the wine stronger and sweeter) into an enamel pot.
2. Add the water.
3. Wash the lemons and oranges and peel them very thinly. Put in the rinds, then add the juice of the fruit.
4. Make a small bag of muslin or any clean white material to hold the spices. You can just put the spices into the must, if you like, but it is better to have them in a bag, since it keeps the wine clearer. Use the spices I list, or others to your taste. Suspend the spice bag in the liquid.
5. Heat the mixture slowly to a boil; keep it simmering for an hour to an hour and a half, clearing away any scum that rises to the top.
6. Let it cool to lukewarm, and prepare the toast and the yeast, dissolved in ½ cup warm water. Put these into the liquid, cover, and set in a warm place (65°-70°).
7. When the wine ceases active fermentation, which should be in ten to twenty-one days, strain, decant, fine if necessary, and bottle.

It is well to let all honey wines stand in bottles lightly corked for quite some time, say eight weeks, before corking them properly. Even then, it is wise to put the wine into heavy bottles and to wire the corks, for now and then they will pop out.

In six months to a year, you will have a splendid, strong, spiced drink.

II

I also made this honey wine, which was strong and dry.

You need: 2¼ gals. water
10 lbs. honey
3 oranges
3 lemons
1 piece toast
½ oz. yeast (2 packages)

1. Put the honey and the fruit, sliced, into the water. Bring to a boil and let it simmer for an hour to an hour and a half, removing any impurities that come to the top.
2. Strain, put the liquid into a crock, and let it cool off.
3. Proceed as in step 6 in the first recipe.
4. When fermentation has ceased, strain, fining if necessary. Bottle in heavy bottles, as in the preceding recipe.

III

This recipe is for a very simple honey wine, one easy to make and a pleasure to drink. It requires the use of dried hops, which are not hard to get. A pound will be enough for many gallons of wine. Hops make the wine clearer and give it a little sharpness.

You need: 2 gals. water
2 oz. dried hops
6 to 8 lbs. honey
3 lemons
3 oranges
1 lb. raisins
½ oz. yeast (2 packages)

1. Put the water, hops, honey, and washed, sliced fruit into an en-

amel pot. Bring to a boil and allow it to simmer for thirty minutes.
2. Strain into a crock. Add the raisins, cut finely, and cool the liquid to tepid.
3. Dissolve the yeast in ½ cup warm water and add it.
4. Cover, stand in a warm place (65°-70°) until fermentation has ended. Strain, decant, and fine if necessary. Don't drink the wine for a year.

Now I shall tell you a little about the different English meads, which were very popular in the Elizabethan days. These recipes are culled from the writers of the sword-buckling days, in what to us are quaint and amusing old herbals, cookbooks, etc. If you are adventurous, you might make them, even as I have done. They are good drinks and will always provoke good conversation.

As I said before, the word hippocras brings bright fancies to the mind, and so I will speak of it first. No, before that, I must tell you of piment or pyment, another kind of mead. It will lead to the first-named.

Let me say here that in every one of these recipes you must add the rind of three lemons and three oranges, well washed, as well as the juice.

Piment

Piment (pyment) is a mead in which wine is used instead of water. It is really a wine mixture, with honey added to grape juice or red wine (preferably). For every gallon of juice or wine, allow about 2 pounds of honey and 2 or 3 sliced lemons and oranges. The addition of the fruit adds piquancy to the flavor. The mixture should rest lightly corked for two or three days before it is strained and bottled. This wine, like the other spiced meads, was usually drunk hot.

Hippocras

Here is a formula of about 1655, showing how W.M., cook to Queen Henrietta Maria, made "hypocras":

"Take four gallons of Claret Wine, eight ounces of Cinamon, three Oranges, of Ginger, Cloves, and Nutmegs a small quantity, Sugar six pounds, three sprigs of Rosemary, bruise all the spices somewhat small, and so put them into the Wine, and keep them close stopped, and often shaked together a day or two, then let it run through a gelly bagge twice or thrice with a quart of new Milk."

These wines were very common until the sixteenth or seventeenth centuries, and were praised and enjoyed by clergy, poets, and knights. So much were they in vogue among the clergy that the church leaders tried again and again to put a stop to their consumption. As early as 817 A.D. there was an order forbidding their use, which of course was ignored.

Sack Mead

This is the same as the first mead (honey wine), except that you use a greater amount of honey—8 pounds to the gallon—which will increase the alcoholic content.

Metheglin

This wine came much later than true mead. During the Middle Ages it was considered a cure for any and every thing. Its very name is derived from the Welsh word *meddyglyn*, meaning "medicine."

Metheglin is a spiced wine and can be made either dry or sweet, depending on the amount of honey used. Four to 5 pounds of honey to a gallon of water will give a drier wine; 6 to 8 pounds to the gallon will increase the sweetness.

1. Put 8 to 10 pounds honey into 2 gallons water.
2. Add spices in a spice bag or put them directly into the liquid: 1 dozen cloves; 1 or 2 sticks cinnamon; 1 ounce ginger; a little balm; angelica; marigold; mace; damask rose—if you have any of them. In fact, put in whatever spices you like.
3. Simmer for thirty minutes, and then strain. If you want the wine strongly flavored, leave in the spices during the period of fermentation.
4. When lukewarm, add ½ ounce yeast dissolved in ½ cup warm water. After the fermentation stops, there is the customary procedure of clearing, fining if necessary, and bottling.

Melomel

This is a mead made of any fruit juice you like, with 4 to 6 pounds honey added per gallon of juice. The process is the same as in the others: simmering, then the addition of yeast. I found one recipe that called for 4 pounds rose hips, boiled and pressed out, for each gallon of water (instead of fruit juice). If you can buy rose hips in an herb store, or if you grow enough roses, these should make a fine and healthful wine. This variant also calls for 4 pounds honey to a gallon of liquid.

Hydromel

1. Into 2 gallons water in an enamel vessel, put: 8 pounds honey; ½ teaspoonful ground cinnamon and a small amount of other

spices you favor; the washed, thin rinds of 4 lemons and 4 oranges; the juice of these citrus fruits.
2. Bring to a boil and simmer for thirty minutes.
3. Strain the liquid into a crock. Cover and put it in a warm place (65°-70°). Fermentation will probably start in a day or two. If it does not, add ½ ounce yeast dissolved in ½ cup warm water.
4. When the fermentation is ended, proceed to clear and bottle.

Put this in strong bottles and let it rest long enough, upright, to make sure all fermentation has stopped.

White Mead

Here is an old recipe, dated about 1720.

"To every gallon of Water add a Pint of Honey and half a Pound Loaf Sugar [white granulated sugar will do as well]; stir in the Whites of Eggs (four), beat to a Froth, and boil it as long as any scum arises. When 'tis cold, work it with Yeast, and to every Gallon put the Juice and Peel of a large Lemon. Stop it up when it has done working and Bottle in ten Days."

This is really very simple and follows the same procedure as the other recipes.

There are other unusual flavors that crown the cup of Anglo-Saxon mead. Adding a quart of rum after fermentation ceases will bring dreams of great cutlass-and-dagger buccaneers roaming the seas for red-haired Queen Bess. Or, if you want dreams of the Far East, add a few drops of almond extract.

Thus you may flavor your honey wines with the tastes of all lands, creating a canticle of drinks, strong and exhilarating.

I will close this honeyed chapter with a recipe for "Dr. King's mead." This was the first mead I made, and it resulted in a royal failure. I found the recipe in *The Queens Closet Opened*. What a title!

Here is how I adapted it, and I shall never know why it went wrong.

For a small amount, take a quart of water, 1 cup of honey, 1 lemon cut in slices, ½ tablespoon of nutmeg. Boil until no scum arises. Add a pinch of salt, the juice of half a lemon, strain and cover. Let it stand until it ferments.

That is exactly what I did, and it ended in a fiercely potent, most unpalatable drink. I never thought of adding sugar syrup. I am certain that would have corrected it. But after that first failure, every mead I made turned out perfectly.

Nut-leaf Wine]

I HAVE INCLUDED THIS WINE in the book for its sheer rarity and also for its unusual bouquet and taste. The leaves of nut trees have a very lovely scent, particularly the young leaves. You can make the wine if you have an English walnut tree, which is somewhat rare up north; or from the leaves of the black walnut, which is common; or from the leaves of the hickory nut tree, which is very common.

The nut tree has had a long career in the lore of many lands. It was an important tree in the Holy Land, and Solomon had his "walnut bower."

As usual, the imaginative Greeks had a tale for the tree. King Dion, of Laconia, had three lovely daughters. When the god Apollo visited the king, his daughters were gracious and hospitable to him. So he rewarded the three maidens with the gift of prophecy, on the condition that they would not misuse the divine gift or try to learn through it knowledge unsuited to women. But alas! Love, in the form

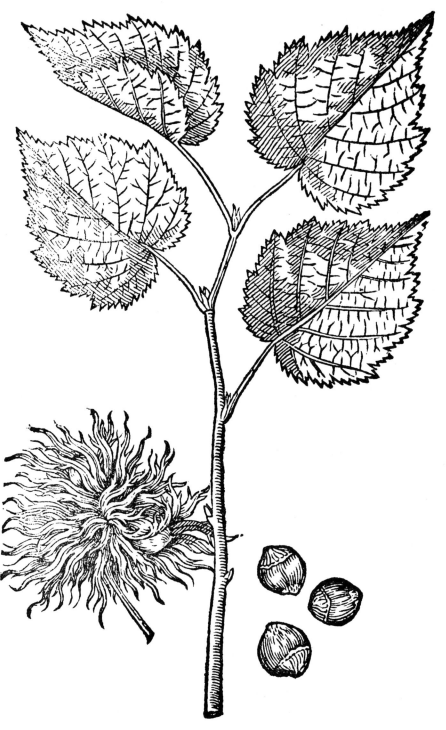

NUT LEAF

of Bacchus, entered their Elysian life, and Carya, the youngest of the three, fell in love with the God of the Vine. The other two, jealous of this love, used their secret knowledge to prevent the two lovers from meeting. Bacchus, in great anger, turned the two jealous sisters into stones and his love into a walnut tree, Nux, from the Latin.

Thus the tree became a symbol of love and the close sisters of love—fecundity, abundance, and regeneration. This belief grew and spread. Both the Athenians and the Romans threw nuts at a bride and groom. Virgil speaks of the custom. In fact, it lived through the ages down to this very day. In Piedmont, Italy, there is a saying: "Bread and nuts are food for married people."

Strangely enough, the walnut tree began to be considered accursed in many parts of Europe, particularly in Italy. Devils and evil spirits were associated with it. In England it was called the witch's tree. Even today it is reputed to produce poison that will kill anything planted under it.

Then came a quirk. It was discovered that the walnut was a remedy for epileptic fits and fevers and would keep lightning from your home and your body. And oh yes, something else! Put a walnut (in Sicily) under the chair where a witch is sitting, and see what happens. She will not be able to rise! What fanciful minds they had in those days! If you beat a walnut long enough, you will beat the devil out of it. And I must not forget the broad-minded Russian proverb: "A dog and a wife and a walnut tree, the longer you beat them, the better they be."

There is a charming legend in Lithuania. When the waters of the Flood began falling on the world, God was eating walnuts. The Lord kept eating the juicy nuts, dropping the shells on the earth. The people of the world tried to flee from the pouring waters, but there was no escape. But the righteous and the just were lucky enough to hold onto the nutshells falling from Heaven. Then a miracle happened. The shells that had been touched by the heavenly hand grew larger—large enough to hold the virtuous—and they climbed in. By the same miraculous divine power, the shells floated right up to Noah's ark, and the good folks climbed into it and so were saved. It is nice to learn that Lithuanian folks just couldn't see all the people of the world drown—only the bad ones.

I heard the preceding story when I was a youngster in Austria —it was the good Christian version. Since then I have learned that the miracle is ascribed to Pekun, the chief god of the ancient Lithuanians, and that only a few friends of Pekun escaped; they then re-peopled the world.

Among some Italians, Saint Agatha is said to have crossed the Mediterranean from Catania to Gallipoli in a walnut shell, and if you rise early in the morning and look out over the sea, you can see that boat even now. Surely the leaves of a tree so steeped in holy lore deserve to be turned into a wine.

The Recipes

I

You need: *4 qts. nut leaves, rather tightly packed down*
2 gals. water
6 to 8 lbs. sugar
2 lemons
2 oranges
1 lb. raisins
½ oz. yeast (2 packages)

1. Pick leaves from a black walnut, English walnut, or hickory nut tree. Be sure they are not wilted and have no sign of scale. Crush them in your hands and put them into a crock.
2. Boil the water and sugar and pour over the leaves.
3. Peel the rinds of well-washed lemons and oranges very thinly and put them into the hot liquid. Cover and let stand for twenty-four hours.

4. At the end of that time, strain the liquid into an enamel pot, squeezing every drop of juice from the leaves. Add the raisins, cut up, and the juice of the lemons and oranges.
5. Dissolve the yeast in ½ cup warm water and pour it into the crock. Then cover and let ferment. This will take at least three weeks, perhaps more.
6. When fermentation ceases, strain the wine into glass jars and let stand (covered, but not closed tightly) for two weeks or more.
7. Now siphon off whatever wine is clear and fine the rest. Bottle and cork.

II

A nice nut-leaf mead can be made in the following manner.
1. Get 2 to 4 quarts clear and healthy nut leaves and put them into 2¼ gallons water in an enamel pot.
2. Add to this 6 pounds honey and boil the liquid for an hour, taking off whatever scum rises to the top.
3. Add yeast and proceed as in the preceding recipe.

Do not drink this wine for at least a year.

[Oak-leaf Wine]

IF YOU HAVE A PLACE in the country where there are oaks—and they are everywhere in the country—you can pick the leaves and easily make an excellent wine with a sherry flavor.

The lore of the "lord of the forest," from which the Greeks

and the Romans and the Scandinavians say man originated, would fill volumes. From the dimmest years, the grandiose oak has been identified with the gods of thunder, lightning, and rain. Lightning came when the gods rubbed two pieces of oak against each other.

The Bible has no end of references to the oak. It is said to have been the tree under which Abraham received the three angels. Underneath an oak Jacob hid the "strange gods" from his children, and the messenger of the Lord that appeared to Gideon sat beneath a great oak. And in an oak Absalom was hanged. There are also several references to it in the Prophets.

The Greeks believed that the oak was the first tree that appeared on the earth. They considered the acorn to be the first food man ate, and so the oak tree was called the "Mother Tree" and became a sacred symbol of worship and of strength. Erisichthon was condemned to eternal hunger for cutting down an oak grove sacred to Ceres. The oak was also sacred to Zeus, who was sheltered by it at his birth. Oak-leaf chaplets were worn at the Eleusinian mysteries.

A Roman who saved the life of another was crowned with oak leaves. Boughs of oak were carried at Roman wedding ceremonies as a sign of fortune and fecundity. I could continue without end telling of the worship and weavings of the giant tree in the life of Hellas and that of the city reared by Romulus and Remus.

The Nordic peoples also invested the oak with mystic and godly powers, and to them, too, it was an object of worship. Great people were buried under it. If a man cut down or maimed an oak, he was punished by the loss of his first-born son.

The Britons dedicated the oak to their god of thunder, and the Druids considered it holy, even as the mistletoe that grew upon it. They performed their sacred rites and human sacrifices under it. And the tree was held in veneration long after the advent of Christianity. The belief in its sacredness, in Christianized form, continues even to this day among many peoples. In Italy, for example, many great miracles happen . . . under the oak tree.

Oaks were also the homes of fairies and dryads and hamadryads, and elves find shelter in them.

Of course, the oak has had its place in medicine. Creeping through an oak cleft will cure hernia and other illnesses. Sleeping under an oak will cure paralysis, and giving oak leaves mingled with oats to a black horse will change him to dapple-gray (this, on the authority of a Danish physician named Paulus). There are many omens attributed to dreaming of the oak, but the omen varies according to the condition of the tree. Thus, to see many oaks means many brave sons. If the tree is withered, it means poverty in old age; but if it has leaves, it means a long and happy life.

With this splendid weight of lore, it is time to turn to a lighter, truly light, theme, for such is oak-leaf wine.

The Recipe

You can pick the leaves of the oak during any period from early spring to late fall. The different leaves will give differently flavored wines. It is interesting to make three wines: one from the spring leaves, one from the full-grown summer leaves, and one when they begin to turn during the fall. Figure on a gallon of leaves for each gallon of wine.

You need: *8 qts. oak leaves*
2 gals. water
6 to 8 lbs. sugar
2 lemons
2 oranges
½ oz. fresh ginger root
1 slice toast
½ oz. yeast (2 packages)

1. Pick the oak leaves, crush them in your hands, and put them into a crock.

2. Boil the water and pour it over the leaves. Cover the crock and let it stand for three days.
3. Strain the liquid into an enamel vessel, squeezing the leaves well, and add the washed, thinly peeled fruit rinds, fruit juice, and ginger. If oranges are not in season, you can substitute lemons for them. You can eliminate the ginger if you do not want the wine to be a little spicy.
4. Bring the liquid to a boil and let it simmer for thirty minutes.
5. Pour it back into the crock. Prepare a slice of toast; cover it on both sides with yeast, dissolved in ½ cup warm water. Add both to the crock.
6. Let the must ferment for ten to twenty-one days (until the visible fermentation ceases), then strain it into glass jars. Now let the wine rest for two weeks or more. If not clear, fine and decant, then put it into bottles and cork.

The wine should not be drunk for nine months or longer.

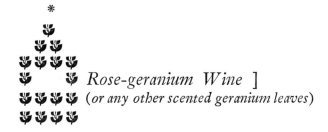

Rose-geranium Wine]
(*or any other scented geranium leaves*)

LET ME RECOMMEND TO YOU a wine made from geranium leaves—rose, spice, or apple geranium. There are over two hundred types of geraniums, each with its own particular scent, and each one can be used for wine making.

In my home, all vinegars are flavored: with rose geranium, tarragon, pineapple-sage, and so on. When used in salad dressings, these vinegars add a delicate and unusual perfume and flavor. With rose-geranium vinegar you will get a perfume and flavor of roses. Put one or two—no more—rose-geranium leaves under the bottom crust of a cake or pie, and you will add a delightful perfume to the food.

It used to be said that rose-geranium leaves were for old-time beaux. Perhaps they rubbed it on their hands and lips to make themselves seem enchanting and romantic. There is the scent of roses in that thought.

The word "geranium" comes from the Latin word *generos*, a crane's bill, because the seed pods resemble the bill of that bird.

There is a flowered Mohammedan legend of how the geranium came to the world. On a warm, sunny day, the Prophet went to bathe in the river and to wash his shirt. When the task was done, he put the shirt on white mallow twigs to dry in the sun. The mallow blushed with joy and was overcome with happiness at the honor. So immeasurably happy was she that she slowly transformed herself into a lovely flower with leaves that were richly perfumed. When the Prophet lifted the dry shirt, behold! Instead of the ordinary white mallow, there was a rich plant with tiny pink flowers and leaves that gave a delicious fragrance—the rose geranium.

Since then, scented geranium leaves have been used to perfume vinegars, wines, cakes and custards, puddings and jellies. It was for that reason I decided to make a rose-geranium wine, for I had not heard of one before, and now I have a dozen bottles of wine giving a perfumed vinous delight.

Let me say here that you can grow the plant on your window sill, just as you can in your country garden, but be sure it has plenty of sunshine. In sunny warmth, it will grow rich enough to give you more leaves than can be used. Most nurseries and flower shops have rose-geranium plants early in the spring, or you can get them from nurseries that specialize in geraniums. And from the leaves you can make a wine that will be delightfully rose-scented and excellent to drink.

The Recipe

You need:
- 4 qts. rose-geranium leaves
- 2 gals. water
- 6 to 8 lbs. sugar
- 4 lemons
- ½ oz. yeast (2 packages)

1. Put the leaves into a crock, crushing them as much as possible with your hands.
2. Boil the water and sugar. Pour the syrup over the geranium leaves.
3. Add the thinly peeled rinds and the juice of the lemons.
4. When the liquid is lukewarm, dissolve the yeast in ½ cup warm water and put it in. Cover and let ferment.
5. When fermentation ceases (it takes from two to three weeks), strain the wine into glass jars and let it stand for another two weeks. If not clear by then, fine, then bottle.

The wine should not be drunk for a year. The longer it matures, the better it will taste.

You can make a wine equally well from any other kind of scented geranium. It will have an exquisite, rare, and unique bouquet and taste.

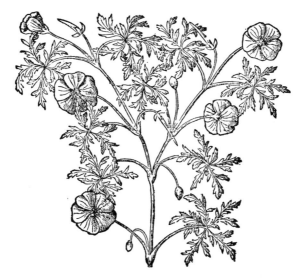

GERANIUM

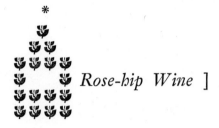 Rose-hip Wine]

THIS IS A WINE said to be unusually good for you and to keep you very healthy. It has millions of vitamin C. I don't know what vitamins really are, but doctors say human beings need them, and I believe and obey. The hips can be secured from your own rose garden or from that of a friend who is so blessed. They can also be bought from commercial herb houses, dried, and probably these would do as well. But I have not tried the purchased dried ones, so I cannot tell. The hips should be gathered when they are red and ripe, which is generally toward the end of high summer.

The Recipe

You need: 2 qts. rose hips
2 gals. water
6 to 8 lbs. sugar
½ oz. yeast (2 packages)

1. Put the rose hips into a crock.
2. Boil the water, in which you have dissolved the sugar, and pour this over the rose hips. When cool, crush the hips with your hands (or a wooden spoon or mallet) as thoroughly as possible.
3. Dissolve the yeast in ½ cup warm water and pour it in. Cover and let stand in a warm place (65°-70°) to ferment.
4. When the fermentation is over, strain and let the wine stand still

ROSE

RUE

for at least two weeks. Then siphon off the clear part, if there is any, and fine the rest.

The longer the wine is allowed to mature, the better it will be. Don't drink it for at least one year.

 Sack Wine]

WHEN I WAS IN ENGLAND, I often drank dry sack sherry and, nosing around among the men selling it and those who knew their wines, I learned that there was a homemade sack wine.

Sack, today, is a dryish wine of the sherry type. But in the olden days it was a name for all strong white wines from Southern Europe. But those good wines, southern or northern, were not for poor folks, so they made their own at home. Thus, a sack wine was created in English homes. I don't know whether the recipe given to me dates back to 1400 or a little later or long before, but that does not matter. Here it is.

You need: *2 gals. water*
*6 to 8 sprigs fresh rue**
a few pinches of fennel seeds
6 lbs. clear honey
½ oz. yeast (2 packages)

* You can grow this easily by buying a plant from one of the herb plant dealers, or you can grow it from seed.

1. Put the water into an enamel vessel, throw in the rue and the fennel, and simmer for about half an hour.
2. Add the honey and simmer for another two hours, removing any impurities that may come to the top.
3. Strain into a crock and add the yeast, which has been dissolved in ½ cup warm water. Cover and set in a warm place (65°-70°) to ferment. This will take from fourteen to twenty-one days. Then strain the wine into glass jars and let it stand for two weeks or more.
4. Siphon off the clear wine and fine the rest before bottling. Let the wine rest for at least one year before drinking. The older it is, the better it will be.

Again and again, I find in old wine recipes the use of honey. It is excellent as a sweetening agent, probably better than sugar.

 Tea Wine]

IF YOU ARE ADVENTUROUS and inquisitive in the realm of folk viniculture and gustatory experiences, make a tea wine.

Of the origin of tea, I know three different stories that make delightful reading. There is a Japanese version, a Chinese version, and an Indian version, all of which are, on the whole, similar.

Daruma, the son of a Hindu king, who became a saint and the founder of the Zen sect (which has suddenly flowered so strongly in our land), retired to Lo Yaeng in China to meditate. There he remained seated for nine years in contemplation and prayer.

One day he accidentally drowsed and a great temptation seized him. While repeating the holy litany, half asleep, he found himself associating the jewel of the lotus of the Sacred Law with that flower worn by a beautiful woman. In his half-sleepy condition, his common human feelings asserted themselves and he approached the lovely vision with rapture and longing. When he awoke (as is ever the sad fortune in such happy, dreaming moments), he found himself beneath the sky "blue as after a rain." He was deeply ashamed, and for his sinful thoughts and desires and in reproach, he cut off his eyelids and cast them on the earth, crying, "Oh, thou imperfectly awakened."

The eyelids transformed themselves into a shrub, the leaves of which resembled the eyelid, and these leaves had the virtue of keeping the body awake. Thus tea came to China.

At the death of the founder of the Zen philosophy, the basis of aesthetic idealism, the disciples gathered before his image and drank tea out of a simple bowl. Out of this Zen ceremony grew the beautifully aesthetic, still semireligious tea ceremony.

The pleasures and ceremonies of tea drinking are said to have started with the legendary Emperor She-nung, "the divine healer," in 2736 B.C. He was a medical man and always boiled his water before drinking it. One day some tea leaves fell into the open kettle, and this started popular tea drinking, first as medicine and then for pleasure.

I had the charmed delight of partaking in a true cha-jin, or tea ceremony, in the beautiful Japanese home of the Ishiharas in Kyoto. Theirs is an old house with a wondrous stone garden and strangely formed trees. There Mr. Ishihara's father had lived, carrying on the august art of block printing, which the son has continued. In this lovely setting the ceremony of the tea was performed by the little daughters of my hostess. It was a perfect wedding of grace, art, beauty, and poetry.

After sitting in the light room for a time, Mrs. Ishihara said in her very quiet voice, "I will show you something."

She led us to the wall of the room, and there raised a small panel. The flickering, shadowy light showed a little miniature chamber with gilded sides. In it were photographs and black-lacquered, upright, narrow wooden slabs with gilded traceries of Japanese lettering—the *ihai*, the mortuary tablets. Before them stood tiny, delicate porcelain cups containing white rice and water.

"This is my parents' shrine. Every day I set before them rice and water, and when I forget, I beg their pardon." Mrs. Ishihara's voice was soft as an afternoon in a shady pine wood.

Since 2736 B.C., tea has not only been drunk, but also sung about in Chinese literature and life, as well as in other lands. Tea is now a word not only used in connection with the tea plant, but with other herbs as well. There are "teas" made from nearly every known herb, and in my own home during the summer, mint tea and pineapple-sage tea are the normal after-dinner drinks.

The Recipes

I

You need: *enough tea leaves to make a gallon or more of tea, any strength you desire*

3 to 4 lbs. sugar
4 lemons
1 lb. raisins
½ oz. yeast (2 packages)

1. Make any amount of tea you desire.
2. Put in 3 to 4 pounds sugar for each gallon and dissolve it completely.
3. Pare the well-washed lemons very thinly; add the rinds and the lemon juice.

4. Take 1 pound raisins for each gallon, cut them fine, and add them.
5. Bring to a boil and simmer for about three minutes.
6. Pour the liquid into a crock; when it is lukewarm, dissolve the yeast in ½ cup warm water and add it to the crock.
7. Cover and let ferment for three days. Then strain off into glass jars, leaving the lids on very loosely so that fermentation will continue. If you want a darker wine, you can let fermentation continue in the crock without straining. Either way, let the wine stand still for at least two weeks or more after fermentation has ceased.
8. Siphon off the clear part and fine any part that is not clear.

Do not drink the wine for at least one year.

II

I have heard that tea wine can also be made without yeast, using exactly the same procedure I outlined, except that no yeast is added. If you make it this way, you must allow the wine to remain in the crock for a full month in a warm place, stirring it now and then and removing any scum that comes to the top. The rest of the procedure is the same as in the first recipe.

This wine also needs at least a year for maturing.

Thus ends my happy pilgrimage of home wine making for folks like you and me. I have attempted to convey to you the pleasure of creating your wines—such as I have when I make mine. And now I sit back smiling warmly, with pleasure and expectation that you will join me in vinous adventures among perfumes of fruits and flowers. They will bring joy to your heart, strength to your body, and inspiration to your mind, even as they have to mine. Amen.

Flowered, Fruited and Flavored

BRANDIES, LIQUEURS, & CORDIALS

BY WAY OF BEGINNING

CORDIALS (FLAVORED BRANDIES) have been made in homes for generations. Recipes for them have been handed down by word of mouth and written word for many, many years. Wherefore they naturally belong in a book of folk drinks. It is pleasant to know that they originated in monasteries with monks who favored and enjoyed good living, and by whom they were legacied into the homes of common folks. Moreover, I have had such excellent success with the unusual after-dinner drinks I have made, and my friends and I have derived so much pleasure drinking them, that I would like you, too, to have the same success and pleasure.

This is important: Cordials are simple to make.

One evening I attended a liqueur-tasting gathering of the Wine and Food Society, and there I drank, for the first time, azari, an orange liqueur. The next day I took a fifth of vodka, the spirit nearest in tastelessness to alcohol, and put into it the well-washed rinds of a few Temple oranges, the juice of the oranges, and a fistful of dried tangerine rind, which I always keep in glass jars for flavoring. Next I added ¼ cup sugar. I let it rest for two months in a dark spot. Then I strained it and put in a few drops of my "many herbs blend" (described a little later), something I do with almost all liqueurs I make, and I had an orange liqueur that can stand up well against azari.

Now, first of all you should know that any spirits can be used to make liqueurs or cordials. If you use pure alcohol (190 proof), you should add one to two glasses of water to decrease its potency. No water should be added to vodka or other spirits. Follow this as a general rule. Whatever spirits you use, I can say with truth that, except for a handful of famous French liqueurs and one Italian liqueur, those made at home can be more delicately varied and better tasting than those you buy. For remember that when *you*, not the factory, make the drink, you can add and subtract ingredients to suit your own taste and fancy. Thus, I have added every possible fruit, flower, or herb to pure alcohol, vodka, marc (flavorless spirits made in France by the peasants), or a cereal (wheat or rice) or potato wine I have made. The liquor will take on the perfume of the plant added, with amazingly pleasurable results.

And so you will find in the following pages: first, unusually flavored and fruited brandies; second, liqueurs or cordials; and finally, unique drinks, little known, but most pleasant to the palate and health, and very easy to make.

Fruits such as cherries, peaches, etc., added to the spirits mentioned will infuse their flavor and perfume into the drink. It is always well to add spices and sugar syrup to the newly created drink.

As I mentioned, I like to use my "many herbs blend" flavoring in practically every cordial I make.

Most liqueurs or cordials have been, and can be, made from synthetic flavors or essential oils, which can be bought in shops that deal with flavors and syrups. The names of some of these firms are listed at the end of this book. You can get oil of peppermint, cloves, anise, and many others, and the same general procedure can be followed with any essence you wish to use. About a dram will flavor one to two quarts of liqueur, depending on your taste.

Mix the essential oil with a quart of any spirit in an openmouthed jar. Then add sugar syrup according to your taste. (You can also add coloring matter if you wish; this can be bought at the same shops that sell the essential oils.) Cover well and let the jar stand in a dark place for one to three months. Now and then, see if any globules of oil have come to the top. These should be taken off with clean blotting or filter paper. Finally, bottle for use.

The strength, bouquet, or flavor of any of the following recipes can be varied as you please to suit your taste and fancy.

Another most pleasurable thought is that your homemade liqueur costs half or less what it would cost to buy it.

No special equipment is required for making brandies or liqueurs: you need only half-gallon or gallon jars with tight covers.

Neutral Spirits

The words "neutral spirits" mean either pure alcohol (190 proof) or any spirits that do not have a pronounced flavor, such as vodka or marc. Remember to reduce the strength of pure alcohol by about one third. In England, gin is often used for making liqueurs. It is as good as any other spirit.

Flaming the Flavor

"*Many Herbs Blend*"

In experimenting with flavoring my liqueurs, I decided that some of them needed a touch of tangy herbs to heighten or sharpen them. After trying varying combinations, I finally worked out one that suits me.

I put a pint of pure alcohol (you can use any neutral spirit) into a glass quart jar. Then I add: 1 teaspoonful dried tarragon; the same amount of caraway seeds and of woodruff; next, I add three pinches of rosemary, ten kernels of allspice, and ten peppercorns. I close the jar and put it in a dark place for one month. After it has stood this long, I use ½ teaspoon or less—sometimes only a few drops—to every liqueur I make. You can vary the amounts of herbs to suit your palate. They give a leap of life to the drink. A pint of "many herbs blend" will last for years.

When you first use it, add only five or six drops, for this "many herbs blend" has a very pronounced taste.

Almond Cordial (*Noyau Cordial*)

Here is a delicious liqueur, not easy to buy, with a unique flavor and perfume.

I

1. Into a gallon jar put:
   2 qts. old rye, bourbon, or vodka
   1 lb. ground sweet almonds
   ½ oz. bitter almond extract (more if you like a stronger flavor)

You cannot buy bitter almonds, and with good reason. They contain prussic (hydrocyanic) acid, and to eliminate it, you must put it through chemical processes too long and too complicated to follow. So, instead, you buy extract of bitter almonds, which is absolutely harmless and will give fine results. Use about 1 ounce, which is equivalent to the oil and flavor you would get from a pound of bitter almonds. Here, again, you can put in more or less to suit your own taste. The sweet almonds, which are the usual almonds you buy, can be ground in a mortar and pestle or pounded with a wooden mallet on a wooden board. It will be easier if you add a little of your spirits while pounding. Or, the almonds can be prepared very easily in a blender. Their brown skin can either be removed or left on. If left on, the liqueur will have a pleasant bittery flavor that I like. If you don't, soak the almonds for about five minutes in a little hot water, and the crinkly skin will come off easily.

Finally, add the thin rinds and juice of 2 lemons, and 3 tablespoons cooled, boiled milk.
2. Cover and stand in a dark place for two months, shaking the jar whenever you think of it. It is best to do this once a day.
3. Strain; add enough sugar syrup to please your taste. Do not throw away the almond paste; it gives a wonderful taste and perfume to desserts, puddings, or cakes.

II

Here is a slightly different way of making your almond liqueur (noyau); this has a somewhat different and very unusual flavor.

Put a quart of brandy or neutral spirits into a half-gallon glass jar and then add ½ handful *each* of prune kernels, apricot kernels, peach kernels, and nectarine kernels (if you can get them).

These kernels should be broken up and pounded a little.

Also add:

3-4 stalks white celery and ½ to 1 ounce bitter-almond extract
1 lb. sweet almonds, pounded
the rind of 1 lemon
about ½ lb. sugar (more if you like it sweeter)

Let this stand for two weeks, shaking it now and then. Then strain. Add to the liqueur ½ glassful of rose water or orange water, then cork. This is a drink with a rare flavor.

Angelica Liqueur

If you grow angelica—and it is easy to grow—it will give you a nice after-dinner drink.

In a wide-mouthed gallon jar put:

1 qt. pure alcohol (190 proof)
1 glass of water (or more, if you wish to decrease the alcoholic

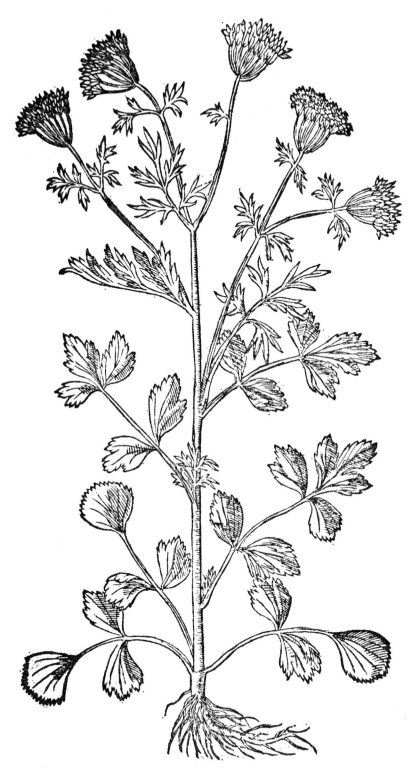

ANISE

strength). Any other neutral spirits, brandy, or whisky can be used—but do not add water to these

½ to 1 qt. angelica leaves, cut with scissors or crushed well by hand
about 1 lb. sugar (more if you like it sweeter)
1 stick cinnamon and 6 cloves

Cover tightly, shaking the mixture a few times. Stand in a cool, dark place for two months. Finally, filter and bottle.

Anisette Liqueur

1. Take from your garden, or buy from a good drugstore or herb store, about 1 ounce green anise seed and ½ ounce coriander seed.
2. Put a quart of whatever spirits you wish into a half-gallon wide-mouthed jar, and then add the anise and coriander seeds. Remember that when pure alcohol (190 proof) is used, it should be diluted with water.
3. Now add a pinch of mace and a pinch of cinnamon.
4. Close the jar well and set it in a dark place for a month to six weeks. Then filter and add sugar syrup to suit your taste. (I use ½ cup white sugar in enough water to dissolve it, and boil it until it bubbles. The boiling sugar-water should be skimmed of any impurities that rise to the top.)
5. Bottle and drink.

The liqueur is also very good if made with anise seed alone.

Arak (Araku) Liqueur

When I was in Polonaruwa, Ceylon, I met a charming lady, Mrs. Miriam Gaskel. I told her that I was interested in traditional native wines, and she told me of an old drink much favored in these parts, which she had made.

It consists of half wild honey and half arak. Arak (native *araku*) is a flavorless spirit fermented from the juice—milk—of the coconut and other palms common in the Far East. It needs flavoring. Of late, I have seen arak (quite high-priced!) in New York liquor shops.

Blackberry Cordial

Get as many blackberries as you can; 4 quarts would be about the right amount. Mash them thoroughly and strain them through cheesecloth or a sieve into an enamel vessel. For each gallon of juice use 1 pound sugar. For each gallon, add 1 teaspoonful cloves and 1 teaspoonful allspice. You can also use cinnamon or any spice you favor. Bring to a boil and simmer for almost ten minutes. Then let it cool off, and when it is quite cool, add a quart of neutral spirits or even good rye or bourbon or brandy. Let the cordial rest for a month before using.

This recipe was given to me by genial Jene Wagner, a seller of fine books in Texas.

Caraway Liqueur (*Kümmel*)

This popular liqueur is easily made.
1. Put a quart of brandy, vodka, or pure alcohol (190 proof) into a half-gallon jar, and add to it 1 ounce caraway seeds. (If you use pure alcohol, dilute it with water to suit your own particular taste.)
2. Let stand for a week, then add a pinch of coriander, a stick of

cinnamon, a bit of orange rind, 3 to 4 kernels of black pepper, and any amount of sugar syrup to suit your taste.
3. Let this stand in a dark place, well covered, for two to four weeks.
4. At the end of this time, clear it through muslin and bottle for use.

Carnation (Gillyflower) Cordial

Gillyflowers, or carnations, were very popular in England for flavoring wines and cordials, and there are many recipes in nearly every book on cookery and medicine. Sack was the favorite wine to which the carnations were added.

The old recipes call for dried carnations or gillyflowers. I used fresh heads of carnations from my garden, and the result was perfect.

First I will give you my own simple recipe, then two old ones. All will give excellent results.

I

Into a gallon jar, put a full quart of any neutral spirits. Cut a quart of highly scented carnation heads and put them in. Close and let stand for six to eight weeks in a dark place. Then strain through cloth, squeezing the juice from the flowers. Add sugar, either syrup or granulated, to your taste, and bottle.

I also added a few drops of my "many herbs blend." You can do likewise if you wish a sharply marked flavor.

II

"Take two ounces of dryed Gilly-flowers [carnations], and put them into a Pottle of Sack, and beat three ounces of Sugar-candy, or fine Sugar, and grinde some Ambergreese, and put it in the bottle and shake it oft, then run it through a gelly bag, and give it for a great

Cordiall after a weeks standing or more. You make Lavender Wine as you doe this." This from *The Queens Closet Opened* (1655).

This is an old and excellent recipe, simple to carry out, and is applicable to any kind of scented herb or flower. White sugar can be substituted for sugar candy, and any spices you like for "ambergreese." In a way, this is quite similar to my own recipe.

III

Sir Kenelm Digby, the home viniculturist and culinary artist—among many other accomplishments—gave this recipe for "Sack with Clove Gillyflowers." He suggested simply adding the flowers to the sack and then waiting until "the Sack hath drawn out all the principal tincture from them, that the flowers begin to look palish."

That is exactly the procedure I have followed with excellent results. However, I added my aforementioned "many herbs blend."

Cherry Bounce

You often hear of this when folks speak of the "good old drinks." Well, it is a good drink today. I have made it, and so can you, for it is very simple to prepare.

I

If you have a wild-cherry tree on your place, pick the fruit before the birds get them. If not, buy the least expensive small black cherries you can find. They are generally better flavored than the large ones.

Cut away half the stems and put the cherries into a gallon or half-gallon glass jar with a wide mouth. Fill the jar about half full. If you buy the cherries, be sure to wash them well first and take off all the stems.

Mash the fruit thoroughly with a wooden spoon, pour over it about half a pound of sugar, and finally fill the rest of the jar with brandy or neutral spirits. Put on the cover, not too tightly, and let stand in a dark place for four weeks. Then mash thoroughly again and let rest for a few months. Most old recipes tell you to strain the brandy at this point. I don't, and my friends and I enjoy the brandy and the pleasant bitter cherries. You can add almonds or spices (any kind of spices you like) to this to give it a spicy flavor.

II

Here is an old Maryland recipe sent to me by Mrs. William Courtland Hart, of Somerset County, Maryland.

"Strain the juice of the cherries through a coarse cloth, then boil it, and put in cinnamon, lemon peel, cloves, allspice, mace, and sugar (you are to be governed in the quantity of each by your taste). Then add 1 gallon of brandy to 4 of juice. At first it will be very strong, but in two months it will lose the strength and it will be necessary to add 1 quart of brandy to every 4 gallons of the bounce."

Of course, this speaks in terms of olden days. This is a fine recipe, one well worth trying.

III

Before closing this, I cannot help but add the words of an old, lovely cherry bounce recipe that begins with "one gallon of good whisky." The recipe is similar to mine, but the ending is a voice from Eden: "This with the addition of ½ gallon of brandy makes a very nice cherry bounce." I am sure it does! Shades of golden days when good whisky and brandy were sold at humane, reasonable prices.

Crème de Cacao

Most women like crème de cacao, so do many men, I among them. The homemade variety costs only one third as much as the purchased, and it is just as good.

1. Buy a pound of whole cocoa beans,* put them in a flat pan, and roast them in the oven. Then grind them in a coffee grinder. If you have no coffee grinder, pound them with any kind of wooden mallet.
2. Put them into a half-gallon or gallon jar and pour in 1 or 1½ quarts of brandy or vodka. Add ½ ounce vanilla beans.
3. Close and let stand for a week in a dark place, then strain.
4. Now boil a pound or two of sugar in half a pint of water for half an hour. When cool, add, according to your taste, to the liqueur.
5. When cool, bottle it.

    It is best not to drink it for at least three months.

Curaçao

Put 1 quart alcohol (or brandy or vodka) into a half-gallon, or better still, a gallon jar. Then add:
the thin rinds of 6 oranges
the thin rind of 1 lemon
3 sticks cinnamon
1 teaspoon coriander seed

---

* You can buy cocoa beans at the Cocoa Exchange in New York. A much simpler way is to buy cocoa extract. Use an amount to suit your particular taste.

Close the jar and set it in a dark, warm place for six weeks. Shake it about once a day. Then filter it through cloth. Next, boil 1 pound sugar in 1 pint of water for about thirty minutes. It should be reduced in volume by about one third. When cool, add it to the liqueur. Bottle.

Many-Flowers Brandy: "The Divine Cordial"

This is a modified and practical recipe of one I found in *The Toilet of Flora*, an old book that fully lives up to its title and is a delight to read. The recipe suggests a kind of continuous process, where you add each flower as it comes in bloom. Thus the cordial is in the making for three seasons: spring, summer, and fall.

Pour 2 quarts brandy into a gallon glass jar. (I used only 3 pints vodka, to allow space for the flowers.) Put in first ½ dram* cloves, the same of mace and orris. (Orris can be bought in any spice store.) Add the thin rinds and juice of two oranges. Now pick a pint of early spring flowers (hyacinths, jonquils) and put them in. Keep the jar well closed and stand it in a dark place. When violets are in season, add a pint of these flowers. When musk roses or any highly scented roses bloom, put in a tight fistful of the petals. As other flowers blossom, keep on adding them: carnations, rosemary, thyme, lavender, sweet marjoram, marigolds, etc. After that, let the jar stand, tightly closed, in a dark place for two to three months. Strain, squeezing the juice out of the flowers. Then add sugar syrup to suit your taste buds. The drink will give superlative satisfaction.

Sometimes this continuous intermarriage of endless flowers and good brandy is called appropriately "the divine cordial," a name fully merited because of its bouquet and flavor.

* 1 dram = ⅛ ounce.

All-Fruit Brandy

Just as you can flavor spirits with many flowers, you can do the same with many fruits, but it is well to use berry fruits and tree fruits separately. Some folks say you can mix them with equally good results. "One shaved his beard, another plucked out the hairs," says the Arabic proverb.

If you use berries, put a pint of every kind you can buy or pick into a glass or earthen jar and set it in boiling water. Let it stay there until those with skins break open. Strain them and add double as much neutral spirits (or any spirits) as there is juice. Add sugar to suit your taste, and let the brandy rest for two to three weeks. Then decant and bottle.

With fruit, put the different fruits—peaches, plums, even pears—all washed, opened, and cut in small pieces, into a gallon jar until it is half full. Fill the rest of the jar with any spirits you choose. Close, and put in a dark place for two months. Then strain and bottle.

Don't throw away the berries or fruit. They are excellent in desserts.

Ginger Liqueur

If you like the taste of ginger, here is a fine ginger liqueur.
1. Put into a half-gallon jar:
    ½ oz. ground ginger or 1 oz. fresh ginger root, well bruised
    ½ lb. cut-up raisins
    2 to 3 washed lemons, sliced

2 oz. macerated sweet almonds and ½ oz. essence of bitter almonds

1 qt. brandy, vodka, whisky, or pure (190 proof), diluted alcohol

2. Finally, add sugar syrup or granulated sugar to suit your taste.
3. Cover and set in a dark place, shaking it often; once a day is best. Continue this for four weeks, then filter and bottle.

Lemon Cordial

A lemon-tasting cordial is a pleasant drink after any meal or on a lovely afternoon, and here is a simple way of making it.

1. Pare thinly one or two dozen good lemons. Put rinds and juice into a gallon glass jar or other vessel that can be closed tightly.
2. For each pint of lemon juice, add half a pound of sugar. Mix well until the sugar dissolves. Then cover and let stand for two days.
3. Put into the fruit juice an equal quantity of white spirits: vodka, or pure alcohol (190 proof) and water. Cover and let stand for a month or two.
4. Finally, strain through a heavy cloth or filter paper. If you don't find the cordial sweet enough, you can add some sugar syrup.

Marigold Cordial

One of my favorite flower wines is marigold wine. The flower is easy to grow and bears well. After trying marigold cordial, I put it in the class of the rose cordials. The bouquet, the perfume, and the flavor are, in their own way, as lovely and as intriguing.

1. Put a quart of alcohol or vodka into an open half-gallon or gallon

GINGER

jar, and then add 1 to 2 quarts marigold flowers. Let the flowers stay in the spirits for two weeks.
2. Strain, and to the flower-impregnated spirits add:
½ pint boiled water, if alcohol is used—no water if vodka is used
sugar syrup to please your taste
about ½ ounce of my "many herbs blend"
3. Cover well and let stand in a dark place for two to four weeks. Then bottle and drink. It has a truly ravishing taste.

Mint Cordial

This is a very popular drink. You can make it with essence of peppermint (don't buy the artificial) or with fresh mint that grows abundantly in any spot where you will permit it to grow.

I

Take a quart of any white spirits and put it into a half-gallon jar. Add as much oil of peppermint as you wish—about ½ ounce is my choice. Add sugar, either in syrup form or granulated, to suit your palate. Shake well every day for a week or two. Then let it rest for two days, and if any oil globules appear on the top, remove them with blotting paper. Bottle, and it is ready to drink.

II

If you use fresh-cut mint from your garden, here is how you make your mint cordial.

Put 1 quart fresh-cut, scented mint leaves and stems into 1 quart neutral spirits in a half-gallon jar, crushing the leaves as you put them in. Cover and let stand in a dark place for two to three weeks. Then strain, getting every drop of juice out of the leaves. Now add sugar to suit your taste buds.

I always put in a few black peppercorns, which sharpen the flavor somewhat. Or you can use a little of my "many herbs blend."

Orange Liqueur

This can be made from oil of orange peel, which can be bought, or from your own fresh orange rinds (oranges must be well washed) or dried tangerine rinds. In my home, the rinds of all tangerines are dried slowly on the stove, put into quart jars, well closed, and kept for flavoring for the whole year.

Put into a wide-mouthed half-gallon or gallon jar:

1 quart pure alcohol (190 proof), vodka, or brandy
a fistful of dried tangerine rinds (I believe they have the strongest amount of essential oil) or the rinds of two or three well-washed fresh oranges

Close the jar and stand it in a dark place for about two weeks or a little more. Then strain, add sugar syrup to please your taste, and a few drops of "many herbs blend." This will add life and zip to the drink. Let stand another two to four weeks, then drink. I feel this is as fine as any curaçao you can buy.

There is another recipe for an orange liqueur on page 362.

Peach Brandy

This is a Maryland recipe that was given to me by Bertha Nathan. It was one used by her mother and grandmother and, I am quite sure, long before that.

"Put 10 pounds peaches in hot lye for a few minutes, then rub off the fuzz. [This really is not necessary; the fuzz does no harm.] Put the peaches into an enamel pot, and put in 7 pounds sugar and a little water, and bring to a boil. When it is cool, pour it into a glass jar, and then pour in 1 quart brandy and 1 pint sugar syrup if it is not sweet enough. Close and put away for two to three months. Then strain and drink it. Use the peaches with dessert."

Peach Liqueur *(or apricot or plum)*

To make 1½ quarts of liqueur:
1. Take 20 peaches, wash them, cut them into quarters and put them into a gallon or half-gallon jar with a wide mouth.
2. Pour ½ pound (or a pound, if you like it quite sweet) of sugar over the peaches; or use one to two glasses of sugar syrup.
3. Put in a stick of cinnamon and 5 peppercorns.
4. Pour over all a quart of neutral spirits or diluted pure alcohol.
5. Cover and set in a dark place for two months. Then strain and bottle. Don't throw away the fruit; use it with desserts.

In each case, a pleasant, nutty, slightly bitter taste can be added to the drink by putting ten to twenty *broken* pits in with the peaches.

Peach, Plum, and Apricot Liqueur: "London's Admirable"

There is an unusually fine cookbook by Margaret Dods called *Cook and Housewife's Manual* (1828), which contains invaluable recipes for fine drinks. Here is one named for a Mr. London, who was renowned as a landscape gardener and author on agricultural subjects. I recommend it highly. It gives a more piquant flavor than the previous recipes.

This one calls for skinning two dozen ripe peaches, quartering them, and removing the stones. Add to this the pulp of two dozen ripe green gages plums, and one dozen ripe magnum plums (yellow plums). For 4 pounds pulp, allow 2 quarts water and 6 pounds sugar. Boil together, simmering for half an hour. Skim and strain. When cool, add 3 quarts brandy or flavorless whisky.

I made mine somewhat differently. I used ½ gallon of peaches, green gages, and dark plums (one third of each fruit) in a little over a pint of water, and added just 1 pound sugar (you can use any amount to suit your taste). Then I added a quart of pure alcohol (190 proof) and 1½ glasses of water to the fruit and let it stand for a week. Then I strained it.

Peach and Rose-Petal Liqueur: "Zahedan Liqueur"

In Zahedan, Persia, where we came down in our plane because of engine trouble, we rested for the night with an Armenian family. I was given a liqueur by my host, which he said he had made, among many others. He graciously gave me the recipe when I asked for it.

Into a gallon jar, put a quart of spirits (he used vodka). Add:

1 lb. honey

2 to 3 handfuls scented rose petals

1 dozen peaches, halved, with their pits (a few of the pits broken open)

Let this stand for two months, shaking it every second day, and then strain. In Zahedan, the rose petals were still in the bottles. It was a rare-flavored drink that went perfectly with fruit and little cups of fresh "Turkish"—here Persian—coffee. I baptized the drink "Zahedan liqueur."

Pineapple Brandy

I like the flavor and taste of pineapple, and so do most folks, wherefore a pineapple brandy or cordial or liqueur is in order. There is nothing simpler to make.

Cut the rind off a pineapple, slice it, and put the pineapple slices into a gallon or half-gallon jar. Pour ¼ to ½ pound sugar on it, and then 1 quart rum, brandy, or neutral spirits. Another way to add the sugar is to place a layer of pineapple, then a layer of sugar over it, and continue this until the pineapple and sugar are used up. Let it stand for about six weeks in the spirits and then strain it. Don't throw away the fruit, for it is excellent in desserts.

If I use neutral spirits. I always add a few drops of the "many herbs blend."

Raspberry Brandy

1. Put 1 quart washed raspberries into a gallon glass jar.
2. Pour over them 1 to 2 quarts brandy. Crush the fruit well with a wooden spoon.
3. Cover and keep it so for ten days, stirring once or twice each day.
4. Then boil ½ pound sugar in as little water as possible. When the syrup is thick and clear, add it to the liquid and fruit. If you like a very sweet product, use more sugar.
5. Stir it again, strain through a cloth, and let it stand for another

week. Then it will need straining once again. If you strain it through filter paper, you will get a clear liquid. When it is finally clear, bottle.

Rose Cordial

Here is a simple rose cordial. The original recipe comes from Miss Leslie's *Directions for Cookery* (1830). I have modified it a little.

1. Take 1 quart rose petals and put them into a gallon glass jar or a crock.
2. Pour over them a little more than 1 quart lukewarm water. Cover and let stand for twenty-four hours.
3. Strain into an enamel or glass vessel, squeezing every drop out of the rose petals.
4. Pick another quart of scented rose petals and put those into the rose water. Let stand for forty-eight hours, then strain, squeezing the petals dry. You can repeat this a few times, as long as you have enough roses, until the liquid is strongly rose-scented.
5. Now add ½ to 1 pound sugar (depending on how sweet you like it), and 1 to 2 quarts of any brandy, vodka, or alcohol (depending on how strong you want it), 1 ounce broken stick cinnamon, and 1 ounce coriander. Cover well and let stand for three to four weeks. Then strain and bottle.

It will be a pleasure to drink.

Rose Liqueur

I am returning to my Zahedan (Persia) adventure. As I mentioned in the "peach and rose-petal liqueur" recipe, we had plane trouble and had to land in the small airfield of that town. An Armenian who was in charge of the field invited us to dinner. After the meal, the host brought out some bottles of liquor, among them a most delightful rose liqueur, and there I learned how to make it. This liqueur should be made *early* in June.

Pick a dozen or two highly scented roses. Pick them early in the morning before the sun has poured its heat over them, for that draws out the perfume. It is also advisable not to pick them the day after a rain.

Separate the petals and remove the white and yellow parts from the ends, the stamen region. Be sure the petals are dry, then put them into a glass half-gallon or gallon jar and pour a quart of neutral spirits over them. Cover well and put in a dark place. Stir once or twice a week for about four weeks.

Then take another dozen scented roses and remove the white and yellow parts from the petals. Dissolve 3 cups sugar in 2 cups water in any enamel pot with a well-fitting cover, and put the rose petals into the liquid. Cover the pot, bring to a boil, then let simmer gently for an hour. Now strain both the rose-petal brandy and the rose-petal syrup into a suitable jar, so that the two blend. Cover the rose liqueur lightly for about twelve hours, then put into bottles and cork well.

This will be a drink that can truly be called nectar.

Scottish Liqueur

I respect and like Scots for their honesty and integrity, and because I have a magnificent godson, Johnny, in that land. Here is a drink befitting the Scots, my godson, and you and me.

Just pour cinnamon essence over a pound (or more) of lump

CINNAMON

STRAWBERRY

sugar. Pour whisky over it and cover. When the sugar is dissolved, bottle.

I made this by filling a quart jar about one third full of sugar, poured ½ ounce cinnamon essence over it, and filled the jar with whisky. It is a magnificent drink. I think leaving five to six cinnamon sticks in the liqueur for a few weeks would do just as well as using the cinnamon essence.

Miss F. M. McNeill describes this drink in her splendid book, *The Scots Cellar*, under the name of "Caledonian liqueur," a nicer title than mine, but I heard of it under the name I call it. Her book, altogether, is a joy to the lover of viniculture.

Strawberry Cordial or a Cordial Water of Sir Walter Raleigh

Just for pleasure and adventure in fine drinking, I will add a cordial that is attributed to the famous, gallant, Elizabethan nobleman—Sir Walter Raleigh.

"Take a Gallon of Strawberries, and put into them a pinte of Aqua vitæ [you can substitute brandy or vodka], let them stand so four or five days, strain them gently out, and sweeten the water as you please, with fine Sugar, or else with perfume." This from *The Queens Closet Opened* (1655) by W. M.; the Queen was Henrietta Maria, wife of Charles I.

When I first made the cordial I used a quart of vodka; later, I used pure alcohol (190 proof), which the vinous-intelligent state of Connecticut permits you to buy. I used only 2 quarts strawberries. But the more berries used, the better the cordial. You can add honey for sweetening. As for perfume, orange water has been suggested. Rose water or any flower perfume (not synthetic) can also be used.

Strawberry Brandy

This good drink is made in the same way as in the preceding recipe.
Put into a gallon or half-gallon jar a layer of strawberries and then a thin layer of white sugar, continuing in this way until the jar is about half full. Then pour over it brandy, pure alcohol (190 proof, diluted one third), vodka, gin, or any grain spirits that do not have a strong flavor. Stand in a warm place for a month or two. At the end of that time, filter and bottle.

Vanilla Liqueur

If you like the taste and scent of vanilla, as most women and many men do, there is a simple way of making the liqueur.

1. Take 1 quart brandy, vodka, or pure alcohol (190 proof, diluted one third) and put it into a half-gallon jar. Then put in two to three sticks of vanilla. Close and let stand for two to three weeks.
2. Boil 1 pound sugar (less, if you don't want it sweet) in 1 pint water until it is clear.
3. Cool and add to the spirits.
4. Let stand three days, then strain and bottle.
    It should rest a month before it is drunk.

UNUSUAL DRINKS

Atholl Brose

My sincere liking and admiration for Scottish folks, and their fine understanding in the realm of "the luscious liquor," prompt me to include one of their famous traditional folk drinks, atholl brose. It was very popular around 1475, but is unquestionably much, much older. There is an excellent report of it in Miss F. M. McNeill's book, of which I have spoken before.

I
  This is the traditional recipe. Mix in a half-gallon jar:
½ lb. clear honey
½ cup cold water
a handful of oatmeal

It will make a thick paste. Then slowly pour 1 quart of good whisky into it. Next, take a silver spoon and stir the mixture until the froth rises. Pour into bottles and cork. Keep it for a few days and then "serve it in a silver bowl."

II

Meg Dods gives this recipe without the oatmeal. Use 1 pound honey and 1 teacupful water. Stir with a silver (or wooden) spoon until well mixed, then add slowly 1½ pints whisky, alias "mountain dew." Stir until it froths, and then bottle. Keep tightly covered.

Whichever way it is made, it is a fine drink.

Bainnecor Liqueur (*Milk Liqueur*)

The Gaelic for "milk" is *bainne*, and *cor* means "heart." This is a liqueur, the heart of which is milk. The name was suggested by Joseph T. Shipley, that most knowledgeable master of words.

The recipe was told to me by my secretary, Marion Duhrels. It was given to her by her mother-in-law, who says that people have used it for "many" years. It is unquestionably the most unusual liqueur recipe here. It is golden yellow in color, with a delicious, indescribably fine flavor.

When I serve it, I generally wager $500 to 10¢ that none of my guests will guess what it is made of. So far, no one has won the bet.

The original recipe calls for: *3 lbs. sugar*
2 qts. Grade A milk
1 qt. pure alcohol (190 proof)
1 lemon
1 oz. vanilla beans

Pour the milk into a gallon jar. Cut the lemon into small pieces (don't waste any of the juice) and add the lemon and juice to the milk. Add the alcohol, then the sugar, then the vanilla. Shake thoroughly and allow to stand for two to three weeks. Filter through filter paper and bottle.

When I made it first, I found that it was a little too sweet, and also that the residue after filtering—a white, creamy curd—was perfectly wonderful when added to any kind of dessert.

On my second trial, I used only ¾ pound sugar to 1½ quarts milk. After filtering, I added a few drops of my "many herbs blend." This gave a little sharpness to the taste.

No other drink has caused so many comments, exclamations of surprise, and so much discussion as this one.

Brown Betty

I had a British Brown Betty! And Lord, how different it is from the American! What distance can do!

1. Into 1 pint water, put 2 good tablespoons brown sugar, 3 slices lemon, a pinch of ground cinnamon, and one or two cloves. Bring to a boil in an enamel pot.
2. In another enamel dish, heat 1 quart strong ale.
3. Then put the two together.
4. Make ½ dozen pieces of toast; if you want to be classic, make them round. Sprinkle the toast with powdered ginger and nutmeg. Float it on top of the liquid and boil for ten to fifteen minutes. Then pour over the whole a full glass of brandy and serve hot. It *will* be!

Bumpo

In Adair, Ireland, the loveliest village in Europe, in a place where for the first time I tasted real poteen, I heard of a drink called "bumpo." By the snakes of Erin, it was a full-sized man's drink! Try it on a cold night or when you have a cold.

Boil 1 pint water with 8 tablespoons brown sugar and a pinch of nutmeg. Pour this into a pint or two of rum. Add the juice of one to two limes, and drink. All cold of any kind will disappear.

Milk Punch

I found an old recipe called "milk punch." This is a good name, for there is a generous punch in it. It is similar to bainnecor.

Put 1 quart rum, brandy, or neutral spirits into a gallon jar; add the thin parings of 6 lemons and 6 oranges, and 1 pound sugar.

Let stand for twenty-four hours in a dark, warm place, shaking it twice a day. Now add 1 quart boiled milk to the spirits. Let stand for one to two days, then strain through filter paper.

Negus

I like the word negus, and I knew it was connected with a drink. Finally I tried it and found it intriguing to the mind and provocative to the spirit. Here is how it is made.

MILK

You can use port, Madeira, or even a cream sherry. I used some of my elderberry wine.

Put into an enamel vessel any amount of any wine you have chosen. Warm it well, but don't boil it. Into 1 quart water, with any amount of brown sugar that pleases your taste buds, put the rind and juice of 1 lemon, 2 to 3 cloves, and cinnamon, ginger, or any spices you like. Bring to a boil and let it boil for a few minutes, then mix it with the wine.

It should be drunk hot. I bottled some of it and reheated it before drinking. It is best when freshly prepared, which you can readily arrange to do before your guests arrive.

Ratafia

In England there is a drink called "ratafia," and it is a good drink, worth making.

Take a half handful of cherry kernels and the same amount of apricot or peach kernels. Pound them to a pulp in a little of whatever spirits you wish to use. Then put the soft pulp into a half-gallon jar. Add ½ pound sugar and 1 quart spirits. Cover and set in a dark place for about eight weeks, shaking it every other day. Then strain, adding sugar if you want it sweeter. Bottle and drink.

Shrub

We have all heard of "shrub," and generally it is connected with a fruit temperance drink, but it can also be an honest, invigorating, even a soothing, drink.

There are innumerable ways of making shrub. Here is a simple recipe.

1. Begin with gentle fruit juices. Put a little less than 1 pint orange juice and the juice of 2 or 3 lemons into a half-gallon or gallon jar. Before squeezing the lemons, peel the rinds very thinly and add them as well.
2. Add 2 quarts rum. Cover and let stand for three days.
3. Put 2 pounds sugar into 1 quart water and bring to a boil.
4. Add the sugared water to the rum and fruit juice, cover, and let stand for two weeks. Then strain and bottle.

This makes a very fine shrub.

Almond shrub

I heard of an almond shrub in England; it is a wonderful drink.

You need: *1 qt. rum or brandy*
1 glass orange juice
rind of 1 lemon
½ to 1 lb. sugar
1 glass milk
5 to 10 drops essence of bitter almonds

Put into a gallon or half-gallon jar the spirits, thin lemon peel, orange juice, and sugar. Now take the essence of bitter almonds (you can use any amount you please) and add it to the glass of milk. Mix thoroughly and pour into the jar. Cover, put in a warm place (65°-70°), and keep it there until the milk has curdled. This takes from one to three hours. Then filter through filter paper and bottle. Rest it for two weeks before drinking it.

Rum shrub

After tasting this once in England, I decided that it was a good man's drink and well worth making. Here is how it is done.

1. Pare 6 oranges very thinly and put the rinds into a gallon jar. Add the juice of the fruit.
2. Add 1 pint cold water, close the jar, and let the liquid stand for three days, stirring it now and then.
3. Strain into an enamel dish, add 2 to 3 pounds sugar, and bring to a boil.
4. Add a full pint of orange juice and strain it again. Finally, add about 2 quarts rum and bottle.

It is a fine drink on a cold day, and you can use it a week after you have made it.

Thandai (from India)

From Bangalore, in the south of India, erudite and charming Mr. Gill gave me a nice native folk drink, common in those parts and drunk by many folks. It is called "thandai."

Indians take 1 pound shelled almonds and ½ pound shelled pistachio nuts, ½ pound muskmelon seeds, and ⅛ to ½ pound cardamon seeds. More seeds of different varieties can be added, according to taste. Grind these together into a thick, wet paste and put them into a jar. Add 1 to 1½ quarts of milk. Shake well for a time, then strain through a cloth. The residue should be reground or repounded finer, and once again milk is added, then strained again through a cloth. This should be done several times, until the nuts are bone dry.

Then sugar is added to taste, and a dash of saffron for color and good smell. Finally, ice is put in, and it is ready to drink.

Adds Mr. Gill charmingly: "Should one want some kind of kick out of it, then add paste of ganja leaves, according to 'capacity.'

"Ganja leaves usually spin one's head pretty fast. With ganja paste in it, we use it for jolly occasions when merrymaking is the main object."

Would you like to know what ganja paste is? It is a preparation made from the fruiting tops of the female ganja plant, a member of the hemp family. The preparation, pulverizing, and paste-making of the leaves is a lengthy ceremony, involving pounding the leaves by foot for days. The result is an intoxicating substance of considerable potency. I believe the process is now supervised by the government, and so I am afraid you'll have to forgo the ganja.

Usquebaugh

Did you ever drink usquebaugh? I did, in an Irish home. And saints, what a drink! I was told one way to make it—there are many. It is worth a trial, for it will warm every cell of your skin and every bone in your body.

You need: 1 qt. good brandy, alcohol, or vodka
            ½ lb. raisins
            ¼ oz. ground nutmeg
            ½ oz. ground cinnamon
            ½ oz. ground cloves
            1 pinch saffron
            ½ lb. brown sugar
            rind of 1 orange

Add all these good things to the spirits, cover, and put in a dark place. Shake it well every day for at least fourteen days; three weeks is even better. Then strain and drink, and you will sing the praises of Ireland, as I do.

White Whey Wine

In Scotland there are many drinks made with milk and wine. Here is one called "white whey," which I drank.
1. Take a quart of milk and bring it to a boil.
2. Put a quart of white wine or sherry or port or Madeira into a gallon jar and add a stick of cinnamon.
3. When the milk is boiling, put it into the wine. Cover and let stand for about two or three hours. Soon the milk will curdle.
4. Let it cool, then strain through filter paper.

The liquid is fine to drink, and the curd is wonderful with desserts.

Selah!

TO HELP YOU

There are organizations, public and private, individuals, and books that will be of help to you in your home wine making. You can go to any of them, as I have, and I am sure they will give you time and advice, as they have given them to me.

Agricultural Colleges and Societies; Garden Clubs and Botanical Gardens

Start off with the agricultural colleges in your state (or province). There are always some folks there who are interested in home viniculture, even as you are.

Thus I found Professor G. Levine, of the Cornell University Agricultural Department, who went to the trouble of preparing some special yeasts for me, to mention just one helpful service. There were many others. I am sure the same holds true of all agricultural colleges—everywhere in the world.

Garden clubs, and botanical gardens and their libraries—in all of them I have always found some people interested and versed in home wine making. Very often someone knew somebody who could give me the necessary help or information. This generally started a chain, leading to the solution.

The Wine Institute of California, 717 Market Street, San Francisco, will also gladly give assistance.

Again, there are large, often national, organizations devoted to home economics, which includes wine making. Particularly in England, there is The National Federation of Women's Institutes of England, 39 Eccleston Street, Victoria, London S.W. 1, which has a division of home wine making. Mrs. I. L. Currie, the Assistant Secretary, gave me a good deal of aid. You can always write them about your problems, and you are certain of sympathy and help.

There are similar organizations in the United States.

Special yeasts

For special yeast you can contact the Berkeley Yeast Laboratory, 3167 College Avenue, Berkeley, California. The director, Mr. J. H. Fessler, is always ready with good advice.

Most British companies that sell home wine-making supplies also have special yeasts. A fine assortment is available from the Grey Owl Research Laboratories, Almondsbury, Gloucester, England.

But if you have neither time nor patience for special yeasts, you can use, as I have indicated in every recipe, ordinary Fleischmann's dry yeast that comes in small quarter-ounce packages. They can be bought in any grocery in America.

General home-wine-making supplies

The Milan Laboratory, 57 Spring Street, New York City 12, has a large assortment of accessories for wine making. They also supply essences and many chemicals that can be used to correct wines.

One of the best companies, with everything necessary for home viniculture, is W. R. Loftus, Ltd., 24 Tottenham Court Road, London W. 1. They ship anywhere in the world.

Barrels can be ordered from the Maslow Cooperage Company, 16-18 Court Street, Brooklyn, N.Y.

Bottles are available from many companies (look in your local classified telephone book under Glass), or from the Anchor Hocking Glass Corporation, 70 Glass Avenue, Lancaster, Ohio. Often they can be secured from a hotel or restaurant.

Corks are sold in many places, such as wine-supply houses or even Woolworth's. A good source is the Dodge Company, Lancaster, Pennsylvania. For champagne corks, I go to the West Company, Phoenixville, Pennsylvania. Mr. Glen W. Graeff, the head of the organization, is also ready with a good deal of helpful advice.

Dried herbs

There are quite a few companies that specialize in dried herbs and leaves. The firm that holds first place is the Greer Drug and Chemical Corporation, Lenoir, North Carolina. Not only do they advertise 186 different herbs, barks, leaves, etc., for sale, but they will collect for you any "American botanical" you desire.

In New York City, Max Van Pels, 114 East Thirty-second Street, can supply many dried herbs.

In most cities there are health stores that sell dried fruits on which no sulphur has been used, special cereals, unblended honeys, and often dried herbs and spices. Many large department stores also have such products.

Fresh herbs

There are many farms and good companies both in this country and abroad from which you can procure herbs to grow. Among these are: The Tool Shed Nursery, Salem Center, New York; Merry Gardens, P.O. Box 68, Camden, Maine; Capriland Herb Farm, North Coventry, Connecticut; and Max Schling Seedsmen, Inc., 538 Madison Avenue, New York, N. Y.

Essences

Fierotti Company, 1420 Lexington Avenue, New York City, carries the natural essences of almost every fruit or flower. These essences and syrups, when added to spirits, make quite good cordials —though personally I prefer natural fruits, flowers, or herbs. Companies that sell wine-making equipment often carry many of these flavors and essences.

Books

The two stores where I have found the best selection of books dealing with my subject were: The Corner Book Shop, 102 Fourth Avenue, New York City; and John Lyle, 24 Burnham Court, Moscow Road, London W. 2. Hatchards, 187 Piccadilly, London W. 1, is also a good source.

Permits

In the United States a permit is necessary to make wine in your home. Any citizen may make two hundred gallons of wine tax-free, provided such wine is for one's personal use. Write for Form 1541 to the Alcohol Tax Unit, Internal Revenue Service, U. S. Treasury Department—either your local branch or the central office in Washington, D.C. There is no charge for this permit.

 GLOSSARY

AROMA

This is really the scent of the fruit of which the wine is made. The older the wine, the stronger the aroma.

ACETIC ACID

The enemy of home wine making. It is the most important component of vinegar. When your wine turns to vinegar, it is full of acetic acid.

BOUQUET

The perfume of the wine. This fragrance comes with long and proper maturing and constitutes one of the fine pleasures of wine drinking.

FINE

To fine wine means to clear it completely of debris, cloudiness, or murkiness. Fining is done by letting the wine rest for a long time, or by means of other agents such as isinglass, eggshells, or commercially bought substances.

FORTIFYING

The process of adding spirits to make the wine stronger or to keep it from changing in the bottle. Fortifying can also be used to stop fermentation.

HYDROMETER

A graduated glass instrument to measure the density of your wine.

ISINGLASS

A thin paper-like substance made from the viscera of certain fish and used in the manufacture of gelatin. The best isinglass comes from sturgeon.

LEES

Sediment found at the bottom of the wine container after fermentation and maturing.

MUST

Any juice in process of fermentation. When it is fermented, it is wine.

Pétillant

The gentle sparkling found in some white wine. This is caused by the carbon dioxide in it.

RACKING

Taking off the clear wine and leaving the lees or residue behind.

TOPPING

Adding wine to keep the bottle properly full.

Vigneron

He, or she, who makes wine.

YEAST

Minute fungi that ferment sugars, producing alcohol in the process.

BIBLIOGRAPHY

Adams, Henry G. *Flowers: Their Moral, Language, and Poetry.* London, H. G. Clarke, 1845.

Albertus Magnus (d. 1280). *The Boke of Secretes.* London, W. Copland, 1565 (?).

Anonymous. *Toilet of Flora.* London, J. Murray and W. Nicoll, 1779.

Apicius. *Cookery and Dining in Imperial Rome.* For the first time rendered into English by J. D. Vehling. Chicago, W. H. Hill, 1936.

Arber, Agnes R. *Herbals, Their Origin and Evolution.* Cambridge, Univ. Press, 1912.

Aylett, Mary. *Country Wines.* London, Odhams, 1953.

Bailey, Liberty H. *The Standard Cyclopedia of Horticulture.* 6 vols. N.Y., Macmillan, 1914-17.

Balfour, John H. *The Plants of the Bible. Trees and Shrubs.* London, T. Nelson, 1857.

Beals, Katharine. *Flower Lore and Legend.* N.Y., H. Holt, 1917.

Beauties Treasury, or the Ladies Vade-Mecum. Being a collection of the newest, most select, and valuable receipts of making all sorts of cosmetick washes, oils, unguents, waters, etc. 1705.

Beeton, Isabella M. *Mrs. Beeton's Book of Household Management.* London, Ward, Lock, 1915.

Bergen, Fanny D. *Animal and Plant Lore.* Boston, Houghton, Mifflin, 1899.

Bertram, James G. *The Language of Flowers: an Alphabet of Floral Emblems.* London, T. Nelson, 1856.

Black, William G. *Folk-medicine.* London, E. Stock, 1883.

Blackwell, Elizabeth. *A Curious Herbal.* London, J. Nourse, 1739.

Calcott, Maria G. *A Scripture Herbal.* London, Longman, Brown, 1842.

Child, Lydia M. *The American Frugal Housewife.* 16th ed. Boston, Carter, Hendus, 1835.

Cowen, Elsa. *Flower Legends.* 1880.

Culpeper, Nicholas. *The English Physitian:* or, An Astrologo-physical Discourse of the Vulgar Herbs of this Nation. London, P. Cole, 1652.

―――― *Simmonite-Culpeper Herbal Remedies.* London, Foulsham, 1957.

Digby, Kenelm. *The Closet of Sir Kenelm Digby, Knight, Opened* (Reprint). Ed. by Anne MacDonell. London, P. L. Warner, 1910.

Dioscorides, Pedanius. *The Greek Herbal of Dioscorides.* Englished by John Goodyer, A.D. 1655 and first printed by Robert T. Gunther, 1933. Oxford, Univ. Press, 1934.

Dods, Margaret (pseud.), *see* Johnstone.

Earle, Alice M. *Old-Time Gardens.* N.Y., Macmillan, 1901.

―――― *Sun-dials and Roses of Yesterday.* N.Y., Macmillan, 1902.

Ellis, Mary. *The Accomplish'd Lady's Delight in Preserving,* etc. 1719.

Evelyn, John. *Acetaria. A Discourse of Sallets.* London, B. Tooke, 1699.

Fernie, William T. *Herbal Simples.* Philadelphia, Boericke & Tafel, 1897.

Folkard, R. *Plant Lore, Legends, and Lyrics.* London, Low, Marston, 1884.

Fox, Helen M. *Gardening with Herbs for Flavor and Fragrance.* N.Y., Macmillan, 1933.

Frazer, James G. *The Golden Bough,* Vols. I, II, III. London, Macmillan, 1918.

Friend, Hilderic. *Flowers and Flower Lore.* 2 vols. London, G. Allen, 1884.

Gerard, John. *The Herball or Generall Historie of Plantes.* London, Iohn Norton, 1597.

Grieve, Maud. *A Modern Herbal.* 2 vols. N.Y., Harcourt, Brace, 1931.

Hardwick, Homer. *Winemaking at Home.* N.Y., W. Funk, 1954.

Harrison, Sarah. *The House-keeper's Pocket-book.* Dublin, E. Exshaw, 1738.

Hollingsworth, Buckner. *Flower Chronicles.* New Brunswick, N.J., Rutgers Univ. Press, 1958.

Ibn Baithar. *Heil- und Nahrungsmittel* aus dem Arabischen übersetzt von Dr. Joseph v. Sontheimer. 2 vols. Stuttgart, Halberger, 1840-42.

Johnstone, Christian Isobel. *The Cook and Housewife's Manual; a Practical System of Modern Cookery and Family Management.* Edinburgh, Oliver and Boyd, 1833.

King, Eleanor A. *Bible Plants for American Gardens*. N.Y., Macmillan, 1941.
Langham, William. *The Garden of Health*. London, 1578 (?).
Leslie, Elizabeth. *Directions for Cookery*. 20th ed. Philadelphia, Carey & Hart, 1844.
Leyel, Hilda W. *The Magic of Herbs*. London, J. Cape, 1926.
M., W. *The Queens Closet Opened*. London, N. Brook, 1655.
Markham, Gervase. *The English House-wife*. 9th ed. London, H. Sawbridge, 1683.
Mason, Charlotte. *The Lady's Assistant* . . . 3rd ed. London, J. Walter, 1777.
McDonald, Donald. *Sweet-scented Flowers and Fragrant Leaves*. London, Low, Marston, 1895.
McNeill, Florence M. *The Scots Cellar*. Edinburgh, R. Paterson, 1956.
Millspaugh, Charles F. *American Medicinal Plants*. 2 vols. N.Y., Boericke and Tafel, 1887.
Moldenke, Harold N. and Alma L. *Plants of the Bible*. Waltham, Mass., Chronica Botanica, 1952.
Moxon, Elizabeth. *English Housewifry*. Leeds, G. Wright, 1764.
National Federation of Women's Institutes. *Home Made Wines, Syrups and Cordials*, ed. by F. W. Beech. London, 1954.
Papyrus Ebers. Translated from the German version by C. P. Bryan. London, G. Bles, 1930.
Paracelsus. *Theory of Signatures: God's Hieroglyphics*.
Parkinson, John. *Theatrum Botanicvm; the Theater of Plants*. London, T. Cotes, 1640.
Physicians of Myddvai, The. Translated by John Pughe. Llandovery, Roderic, 1861.
Plat, Hugh. *Delightes for Ladies*. London, Hvmfrey Lownes, 1602.
Pliny the elder. *Natural History*. Translated by John Bostock. 6 vols. London, 1855-57.
Rohde, Eleanour S. *A Garden of Herbs*. Boston, Medici Society of America, 1921.
——— *Old English Herbals*. N.Y., Longmans, Green, 1922.
Rothery, Agnes E. *The Joyful Gardener*. N.Y., Dodd, Mead, 1949.
Shepherd, Roy E. *History of the Rose*. N.Y., Macmillan, 1954.
Singer, Charles J. *Early English Magic and Medicine*. London, British Academy (Proceedings), 1920.
Skinner, Charles M. *Myths and Legends of Flowers, Trees, Fruits and Plants*. Philadelphia, Lippincott, 1925.
Step, Edward. *Herbs of Healing*. London, Hutchinson, 1926.
Theophrastus. *Enquiry into Plants, and Minor Works of Odours and Weather Signs*, with an English translation by Sir Arthur Hort. 2 vols. London, W. Heinemann, 1916.
Thiselton-Dyer, Thomas F. *The Folk-lore of Plants*. London, Chatto and Windus, 1889.

Thompson, Charles J. S. *The Mystery and Lure of Perfume.* Philadelphia, Lippincott, 1927.
Tritton, Suzanne M. *Amateur Wine Making.* London, Faber and Faber, 1956.
Turner, William. *A New Herball.* 2 vols. London, S. Mierdman, 1551-62.
Walsh, James J. *Medieval Medicine.* N.Y., Macmillan, 1920.
Warnke, F. *Pflanzen in Sitte, Sage und Geschichte.* 1878.
Wilson, A. M. *The Wines of the Bible;* an examination and refutation of the unfermented wine theory. London, Hamilton, Adams, 1877.
Wirt, Elizabeth W. *Flora's Dictionary.* Baltimore, F. Lucas, 1830.
Woolley, Hannah. *The Queen-like Closet; or, Rich Cabinet.* 5th ed. London, R. Chiswel, 1684.
Wright, Helen S. *Old-Time Recipes for Home Made Wines, Cordials and Liqueurs.* Boston, Page, 1922.

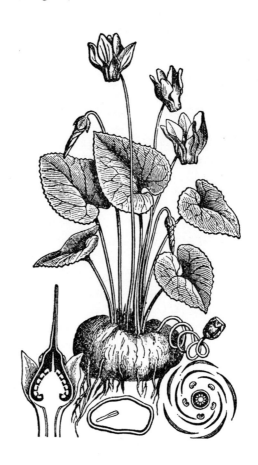

INDEX

Aids in home wine making, 399-403
All-fruit brandy, 375
Almond cordial, 365-66
Almond shrub, 395
Almond wine, 321-24
Angelica liqueur, 366-68
Angelica wine, 251-54
Anisette liqueur, 368
Apple butter, 88-89
Apple pie, 88
Apple wine, 83-94
Apricot liqueur, 380
Apricot wine, 95-97
Arak (araku) liqueur, 368-69
Atholl brose, 389-90
Azari (orange liqueur), 362

Bainnecor liqueur, 390-91

Balm wine, 254-58
Banana wine, 98-100
Barley wine, 215-18
Bees wine, 325-26
Beet-leaf wine, 220-21
Beetroot wine, 218-20
Berry wines, 283-318
"Bierre douce" (pineapple wine), 135
Blackberry cordial, 369
Blackberry wine, 285-91
Blueberry wine, 291-93
Books on wine making, 402-403
Bottles, 68-69, 401
Brandies
  all-fruit, 375
  many-flowers, 374
  peach, 379-80
  pineapple, 382

Brandies (*con't.*)
    raspberry, 382-83
    strawberry, 386
Brandies and liqueurs, preparation of, 361-64
Brandy, use of, in wine making, 52, 59, 63-64
Brandy wine, 327-28
Bread wine, 328-29
Brown Betty, 391
Bumpo, 392

Cantaloupe (muskmelon) wine, 120-22
Caraway liqueur (kümmel), 369-70
Caraway wine, 258-60
Carnation (gillyflower) cordial, 370-71
Carnation (gillyflower) wine, 161-63
Carrot "sippet," 222
Carrot wine, 221-24
Celery wine, 224-25
Cereal and vegetable wines, 213-48
Cereals, use of, in wine making, 58, 59
Cherry bounce, 371-72
Cherry wine, 100-104
Chickweed wine, 261-62
Cider wine and champagne, 332-33
Citrus, use of, in wine making, 59-60
Clove wine, 262-66
Clover-flower wine, 164-66
Cordial water of Sir Walter Raleigh, 385
Cordials, *see* Liqueurs
Corks, 69-70, 401
Cowslip wine, 166-71
Cranberry wine, 293-94
Crème de cacao, 373
Curaçao, 373-74
Currant wine, 294-96

Daisy wine, 172-76
Dandelion wine, 177-84
Date wine, 104-108
Dill wine, 266-68
"Divine cordial," 374
Dumplings, sweet meat, 193

Elder-flower wine, 185-88
Elderberry-raisin wine, 154

Elderberry wine, 296-302
Essences, *see* Oils, essential

Flower wines, 159-212
Flowers, many-, brandy, 374
Fruit, all-, brandy, 375
Fruit wines, 81-158

Geranium-leaf wine, 349-51
Gillyflower (carnation) cordial, 370-71
Gillyflower (carnation) wine, 161-63
Ginger liqueur, 375-76
Ginger wine, 268-70
Goldenrod wine, 189-90
Gome (sugar syrup), 53-54
Gooseberry wine, 303-305
Grape-leaf wine, 113-14
Grape wine, 108-13
Grapefruit wine, 114-16
"Green pudding" (sweet meat dumplings), 193

Herb soup, 193
Herb wines, 249-82
Herbs, *see* Spices and herbs
Hippocras, 339
Honey wine (mead), 333-42
Huckleberry wine, 305-306
Hydromel, 340-41

Isinglass, use of, in wine making, 64-65

Kümmel, 369-70
Kwas, 329-30

Lemon cordial, 376
Lemon wine, 118-20
Liqueurs
    almond cordial, 365-66
    angelica liqueur, 366-68
    anisette liqueur, 368
    apricot liqueur, 380
    arak (araku) liqueur, 368-69
    bainnecor liqueur, 390-91
    blackberry cordial, 369
    caraway liqueur, 369-70
    carnation (gillyflower) cordial, 370-371

Liqueurs (*con't.*)
  cherry bounce, 371-72
  cordial water of Sir Walter Raleigh, 385
  crème de cacao, 373
  Curaçao, 373-74
  "divine cordial," 374
  ginger liqueur, 375-76
  kümmel, 369-70
  lemon cordial, 376
  "London's admirable," 380-81
  marigold cordial, 376-78
  milk liqueur, 390-91
  mint cordial, 378-79
  noyau cordial, 365-66
  orange liqueur, 362, 379
  peach and rose-petal liqueur, 381-82
  peach liqueur, 380
  peach, plum, and apricot liqueur, 380-381
  plum liqueur, 380
  rose cordial, 383
  rose liqueur, 383-84
  rose-petal and peach liqueur, 381-82
  Scottish liqueur, 384-85
  strawberry cordial, 385
  vanilla liqueur, 386
  "Zahedan liqueur," 381-82
"London's admirable" (peach, plum, and apricot liqueur), 380-81

Many-flowers brandy, 374
"Many herbs blend," 364
Marigold cordial, 376-78
Marigold wine, 190-96
May wine, 270-71
Mead (honey wine), 333-42
  nut-leaf mead, 346
  sack mead, 339
  white mead, 341-42
Melomel, 340
Melon wine, 120-22
Metheglin, 339-40
Milk liqueur, 390-91
Milk punch, 392
Mint cordial, 378-79
Mint wine, 272-74
Mulberry wine, 307-11

Muskmelon wine, 120-22

Negus, 392-94
Noyau cordial (almond cordial), 365-366
Nut-leaf wine, 342-46

Oak-leaf wine, 346-49
Oils, essential, 363; suppliers of, 402
Onion wine, 226-29
Orange liqueur, 362, 379
Orange wine, 122-26
Organizations helpful in wine making, 399-400

Pansy wine, 196-201
Parsley–pineapple-sage–mint wine, 279-280
Parsley wine, 276-79
Parsnip wine, 230-32
Parsnips au Melita, 230
Peach and rose-petal liqueur, 381-82
Peach brandy, 379-80
Peach liqueur, 380
Peach, plum, and apricot liqueur, 380-381
Peach wine, 126-30
Pear wine, 130-31
Permit for wine making, 403
Piment, 338-39
Pineapple brandy, 382
Pineapple-sage wine, 282
Pineapple wine, 132-35
Plum liqueur, 380
Plum wine, 136-41
Pomegranate wine, 141-44
Potato wine, 232-36
Potatoes, use of, in wine making, 58, 59
Prune wine, 145-47
Punch, milk, 392

Quince wine, 147-50

Raisin-elderberry wine, 154
Raisin wine, 151-54
Raisins, use of, in wine making, 56, 58
Raspberry brandy, 382-83
Raspberry wine, 311-14

Ratafia, 394
Rhubarb wine, 155-58
Rice wine, 238-42
Rose cordial, 383
Rose-geranium wine, 349-51
Rose-hip wine, 352-53
Rose liqueur, 383-84
Rose-petal and peach liqueur, 381-82
Rose wine, 201-12
Rum shrub, 396

Sack wine, 353-54
Sage wine, 280-82
Sake (rice wine), 238-42
Scottish liqueur, 384-85
Shrub, 394-96
Spices and herbs, use of, 57-58, 362-63, 364; suppliers of, 401-402
Spinach wine, 242-43
Spirits, neutral, 362-63
Strawberry brandy, 386
Strawberry cordial, 385
Strawberry wine, 315-18
Sugar, in wine making, 46-47, 51, 52, 53; see also Wine making: sweetness
Sugar syrup, 53-54, 362-63

Tea wine, 354-57
Thandai, 396-97
Toast, use of, in wine making, 50
Tomato wine, 244-47

Usquebaugh, 397-98

Vanilla liqueur, 386
Vegetable (and cereal) wines, 213-48
Vinegar, formation of, 58-59

Weina palama (plum wine), 139
Wheat wine, 247-48

White whey wine, 398
Wine glasses, 79-80
Wine making
  aids, 399-403
  cost, 31-33, 36, 363
  ingredients, handling of, 42
  labeling, 71-72
  permit, 403
  process
    alcoholic content, see strength, below
    bottling, 68-71
    clearing, see straining, below
    coloring, 60
    corking, 69-71
    fermentation, 45-54, 67
      curtailment of, 53, 67
    fining, see straining, below
    maturing and storing, 52, 73-77
    racking, 66
    siphoning, 64, 66, 69
    straining and fining, 52, 62-66
    strength, 47, 51, 59, 60-61, 63
    sweetness, 52, 53-56, 58, 78, 362-63
    taste and flavor, 51, 52, 57-58; see also maturing, above
    temperature, 52, 63
    topping, 65-66
    vinegar, formation of, 58-59
  quantity, 43-44
  record book, 41-42, 72, 77
  season, 33-34
  serving, 77-80
  supplies, 36, 48, 401-402
  utensils, 35-40, 401

Yeast, 32, 47-50, 400-401; see also Wine making: fermentation

"Zahedan liqueur" (peach and rose-petal), 381-82